# Contents

**PLACES AND CASES**

# Physical Geography and People

## Peter Webber and Neil Punnett

### Series Editor
### Peter Webber

**Stanley Thornes (Publishers) Ltd**

Text © Peter Webber and Neil Punnett 1999

Original line illustrations © Stanley Thornes (Publishers) Ltd 1999

The right of Peter Webber and Neil Punnett to be identified as authors of this work has been asserted by them in accordance with the Copyright, Designs and Patents Act 1988.

First published in 1999 by:
Stanley Thornes (Publishers) Ltd
Ellenborough House
Wellington Street
CHELTENHAM GL50 1YW
England

99 00 01 02 03/ 10 9 8 7 6 5 4 3 2 1

A catalogue record for this book is available from the British Library.

ISBN 0-7487-4303-0


Designed by Peter Tucker, Holbrook Design Oxford Ltd

Typesetting and layouts by Hardlines, Charlbury, Oxford

Illustrated by Hardlines, Nick Hawken, Steve Smith

Cover photo by permission of Science Photo Library

Picture research by Penni Bickle

Reprographics by Blenheim Colour Ltd

Printed and bound in China by Dah Hua Printing Press Company Ltd

## Acknowledgements

With thanks to the following for permission to reproduce photographs and other copyright material in this book:

Aerofilms, 70G, 80A; Associated Press, 14B, 6A; British Geological Survey, 19F; The *Citizen*, Gloucester, 77D; Coventry Evening Telegraph, 66C; John Edwards, 66A, 66B, 67C, 67D; Frank Lane Picture Agency, 94A, 94B; Geoscience Features, 75B, 89C, 89D; Guy Murray, 85D, 85E, 86F, 86G; NERC Satellite Station, Dundee, 23C; Nick Groves, 5 (top), 20H, 20G; Northern News and Picture Agency, 69F; Jocelyn Pritchard, 8C, 84A, 84B, 87A, 87B, 85C, 86H; Neil Punnett, 49J, 50G-J, 57B, 58A-C; Science Photo Library, 74-75, 76A; Scotland In Focus, 8C, 9 (all); Still Pictures, 41B, 43C, 43D, 55D, Mark Specht/StillPictures, 63A, 91B; Surrey Evening Herald, 28D; Tony Waltham/Geophotos, 44F, 69E, 89E (middle and bottom left), 93E; Daniel Webber, 5 (bottom), 60B, 60D; Mark Webber, 38D; Peter Webber, 34E, 34F, 38E, 39F, 39G, 39H, 68B, 69D (2), 71, 72, 73H, 73I, 89E (right).

East Riding of Yorkshire Council, 73J; The *Guardian*, 63B, 71B; The *Independent*, 54A; Kompass, Innsbruck, 60C (Map No. 88, Monte Rosa), 61E (Map No. 97, Omegna-Varallo-Lago d'Orta); The *Orlando Sentinel*, Florida, 25A, 26B, 26D; © Telegraph Group Ltd, London, 1999, 71A; Touring Club Italiano, Milan, Italy, 61F.

Map extracts 38A, 48D, 72E, 92A are reproduced from Ordnance Survey 1: 50,000 (92A) and 1: 25,000 scale mapping with permission of Her Majesty's Stationery Office © Crown Copyright. Licence Number 07000U.

Every effort has been made to contact copyright holders. The publishers apologise to anyone whose rights have been overlooked, and will be happy to rectify any errors or omissions at the earliest opportunity.

# Introduction

## To the student

This book about Physical Geography is one in a series of five textbooks for GCSE Geography. The other books cover the United Kingdom, Europe, the World and Environments.

You will find that the majority of the book consists of case studies. There is some background information about a topic before many of the case studies are introduced. Glacial processes, for example, are introduced on pages 78–79 before a detailed study of the Rhône Glacier and glaciated landscape in the English Lake District. However, if you are going to make the best use of a case study, you need to have some background knowledge. It is assumed that you use a core geography textbook and will have some class time to make sure you know the definitions and the answers to any questions in the 'Do you know?' boxes that introduce topics. This case study approach allows you to broaden and deepen your knowledge and understanding.

The case studies have been carefully chosen to cover the main topics you need for your GCSE syllabuses. So you will find, for example, case studies on earthquakes, volcanoes, tornadoes, rivers, floods, coasts and glacial landforms. There are also studies on climate types and vegetation systems.

Most GCSE examinations either include case studies for you to analyse, or ask you to use a named example you have studied. This book, therefore, gives you practice and examples. You will find that the activities throughout the book will help you develop the different skills you will need in examinations. These include using photographs, tables, graphs, maps, diagrams and charts, as well as reading sections of text and completing decision making exercises. This book gives you plenty of practice!

The symbol ➼ suggests that you write at greater length and in more detail. Your answer should be at least a paragraph in length.

Some of the words which appear in **bold** throughout the book are key terms which are defined in the glossary on pages 95–96.

Geography is all about how the world works – the natural world and the human world – and is about more than just examinations. So we hope this book will help you take an interest in, and begin to understand the world around you.

Enjoy your Geography!

Devastation in Plymouth, Montserrat, 1997

Coastal erosion in Norfolk, UK

# Location of case studies

**Sweden and Finland** Coniferous forest page 41

**Eastern Europe** Floods pages 64–65

**Rhône glacier** pages 50–51

**N Italy** Rivers pages 59–61

**UK** Case studies listed page 8

**Monument Valley, USA** Deserts and slopes pages 90–91

**S Italy** Earthquake page 15

**Florida** Tornadoes pages 25–26

**Black Sea** page 75

**Montserrat** Volcanoes pages 16–20

**Amazon Basin, Brazil** Rainforest pages 31–32

**The Gambia** Savanna pages 33–34

**Figure A: The location of case studies**

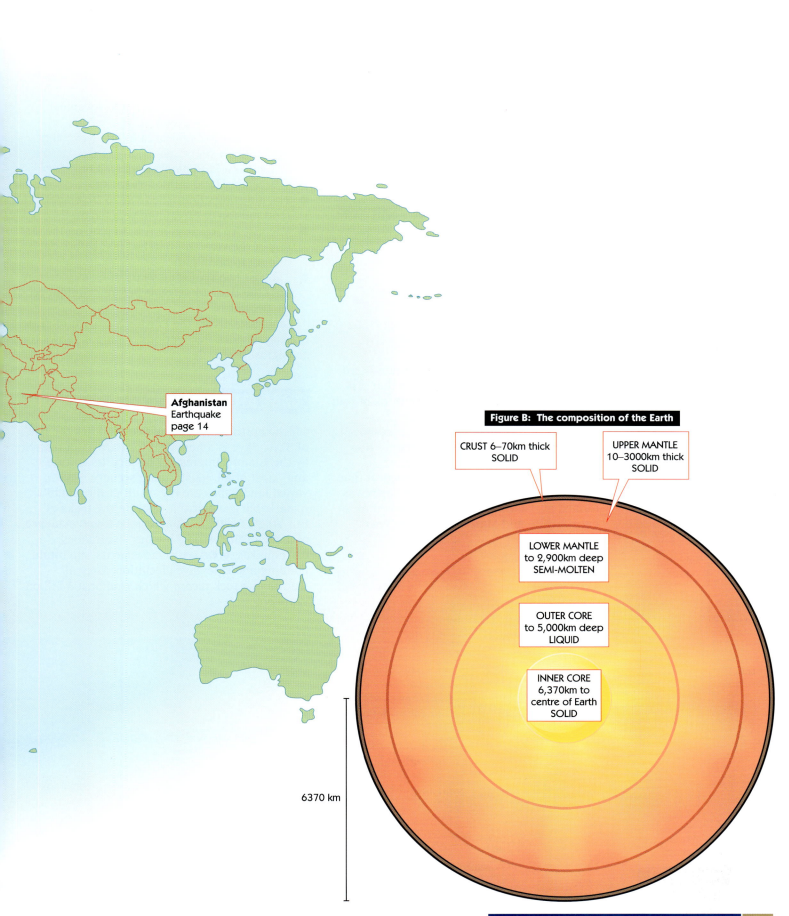

**Afghanistan**
Earthquake
page 14

Figure B:  The composition of the Earth

CRUST 6–70km thick
SOLID

UPPER MANTLE
10–3000km thick
SOLID

LOWER MANTLE
to 2,900km deep
SEMI-MOLTEN

OUTER CORE
to 5,000km deep
LIQUID

INNER CORE
6,370km to
centre of Earth
SOLID

6370 km

# 1

# Geological time

## Main activity

A case study of the Isle of Skye illustrates the concept of geological time. Activities include drawing a time line and writing an introduction to Skye's scenery for visitors.

## Key ideas

● The Earth is 4.7 billion (4,700,000,000) years old.
● The Earth is composed of layers of solids and liquids with a thin crust on the surface.
● Oceanic crust is more dense than continental crust.
● The UK has a wide variety of landscapes dating from different eras of the Earth's history.

It is thought that our planet is 4.7 billion years old. This is such a vast period of time that it is difficult for us to comprehend. If we think of it in terms of one year it is rather easier.
● Each day of this imaginary year represents 13 million years. At that scale, a person's life would last less than half a second.
● From the first billion years of geological time little survives. Assuming that the Earth was formed on 1 January, the oldest known rocks date back to mid-March (3.7 billion years ago).

## ▼ Questions

1  How old is the Earth thought to be?
2  Study Figure A and the text.
   a  In which geological eras did the following events occur?
      (i)    the first animals appeared
      (ii)   the earliest fish appeared
      (iii)  the earliest land plants appeared.
   b  in which geological periods did the following events occur?
      (i)    the appearance of the earliest fish
      (ii)   the appearance of the earliest land plants
      (iii)  the dinosaurs flourished.

## Do you know?

? The Isle of Skye has some of the oldest rocks in Europe.

● The first forms of life appeared in the seas in mid-May (3 billion years ago).
● Animals first appeared in early November (800 million years ago).
● The earliest fish appeared in late November (420 million years ago).
● The earliest land plants appeared on 1 December (390 million years ago).
● The dinosaurs flourished 18–26 December (150 million to 60 million years ago).
● The last Ice Age began at 8.30 p.m. on the last day of the year and finished just before one minute to midnight.
● The last two thousand years of history occupy only the final 14 seconds before midnight on 31 December.

Geologists divide the Earth's history into **eras**. Eras are divided into **periods**. Figure A shows the geological timescale.

## CASE STUDY: The Isle of Skye

Off the north-west coast of Scotland lies the Isle of Skye. Skye is a honeypot for tourists, attracted by the island's interesting history, its culture and, most of all, its stunning scenery. Skye's varied landscapes are the result of the island's remarkable geological history. In the rocks of this single Scottish island can be found evidence of much of the world's history – volcanoes, earthquakes, great earth movements, mountain building and extensive glaciation. They all feature in a rock record which is almost 3 billion years old.

Figure C: Lewisian gneiss

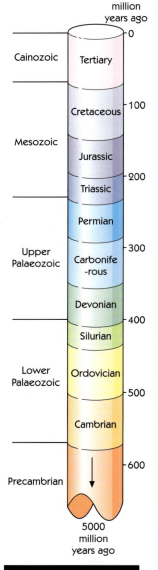

million years ago

Cainozoic — Tertiary — 0

Mesozoic — Cretaceous — 100
Jurassic
Triassic — 200

Upper Palaeozoic — Permian
Carbonife-rous — 300
Devonian — 400
Silurian

Lower Palaeozoic — Ordovician — 500
Cambrian — 600

Precambrian

5000 million years ago

**Figure A: The geological timescale**

# The geological history of the Isle of Skye

Skye has some of the oldest rocks in Europe, dating back almost 3 billion years.

**3 billion years ago** The oldest rocks are found on the Sleat peninsula in the south of the island. These ancient rocks are called Lewisian gneiss (Figure C). Since their formation these rocks have been broken and folded, stretched and compressed by great earth movements. Today they form low, rounded rocky hills.

**1.1 billion years ago** Fast flowing rivers deposited a great depth of pebbles and grits which became **Torridonian sandstone** (Figure B).

**550 million years ago** Sands, silts and muds were deposited beneath a shallow sea. These deposits became Cambrian and Ordovician sandstones and limestones. The Caledonian Mountain Belt was formed at this time.

**230 million years ago** Skye was now part of a hot, dry desert area. Sand and mud were laid down in shallow warm seas which teemed with life. Fossil-rich sedimentary rocks were formed including Triassic and Jurassic limestones, sandstones, shales and conglomerates.

**140 million years ago** At the end of the Jurassic period, a thick layer of mud was deposited which became upper Jurassic shales.

**95 million years ago** During the Cretaceous period, Skye was covered by an extensive sea in which shells and bones of marine life built up to form chalk. Later erosion has removed almost all the Cretaceous rocks from Skye.

**65 million years ago** Skye was raised above sea level. Greenland began to split away from North West Europe as the Atlantic Ocean started to form. Magma erupted building up the vast lava plateau of northern Skye (Figure D).

**60 million years ago** Large volcanoes developed. The cones have been removed by erosion, but the roots of the volcanoes remain as the granite of the Red Hills (Figure E) and the **gabbro** of the Cuillin (Figure F). These highly resistant rocks form some of the most dramatic mountains in the UK.

**2 million years ago** The climate started to cool and a series of glacial advances and retreats began. Most of the soils and recent sedimentary rocks were eroded away by the ice and, about 450,000 years ago, Skye became an island.

**10,000 years ago** The climate warmed rapidly and the last glaciers melted, leaving very clear evidence of their presence. The landforms of glacial erosion and deposition have been slowly modified by weathering and erosion, but remain amongst the best preserved examples in the UK (Figure G).

Figure D: The lava plateau

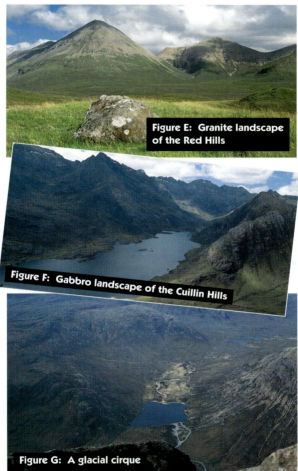
Figure E: Granite landscape of the Red Hills

Figure F: Gabbro landscape of the Cuillin Hills

Figure G: A glacial cirque

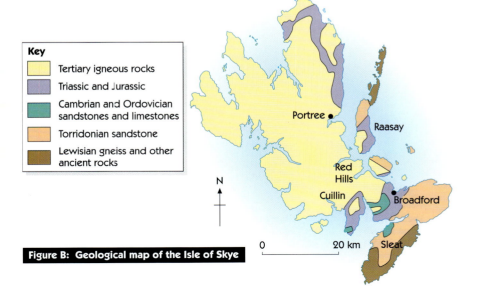

**Key**
- Tertiary igneous rocks
- Triassic and Jurassic
- Cambrian and Ordovician sandstones and limestones
- Torridonian sandstone
- Lewisian gneiss and other ancient rocks

Portree
Raasay
Red Hills
Cuillin
Broadford
N
0    20 km
Sleat

**Figure B: Geological map of the Isle of Skye**

## ▼ Questions

**1** a  Where is the Isle of Skye?
  b  To which group of islands does Skye belong?

**2** a  Name Skye's oldest rocks.
  b  In which part of the island are they found?

**3** In which geological periods did the following events occur on Skye?
  a  Skye was part of a hot, dry desert area
  b  chalk formed
  c  large volcanoes developed
  d  the Ice Age.

**4** Draw a time line to show the main events in Skye's geological history. Let 1 cm represent 100 million years.

**5** Write the introduction to a guide book for visitors to Skye explaining why the island has such varied and dramatic scenery.

# The physical background of the UK

Figure A is a relief map of the UK showing major rivers and the location of the case studies in this book. Some case studies are introduced as 'It happened in the UK'. The imaginary Tees–Exe line divides the country, with older rocks to the north and younger rocks to the south (Figure B).

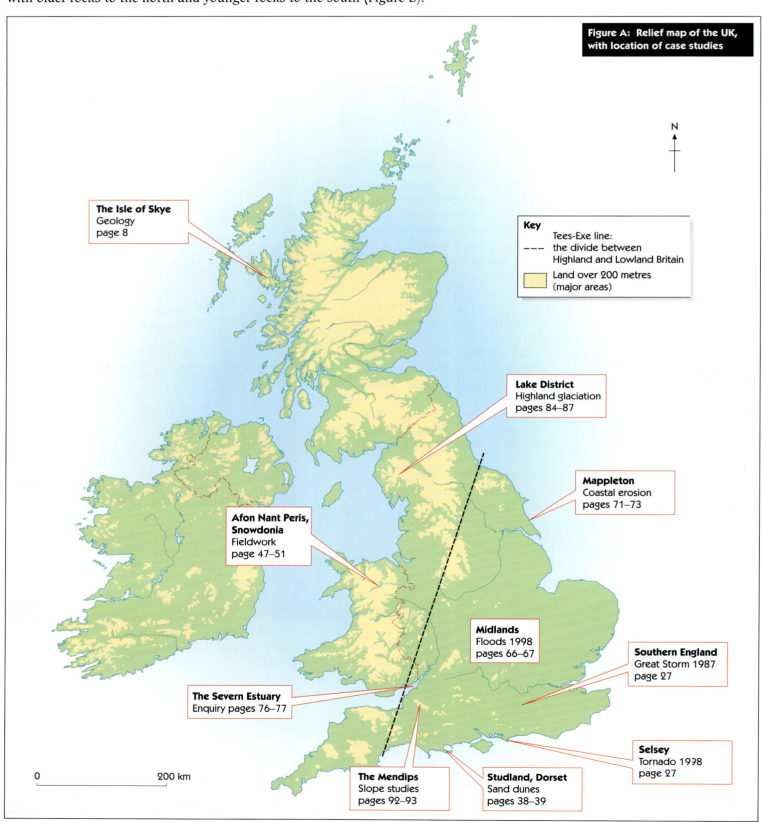

**Figure A: Relief map of the UK, with location of case studies**

N

**The Isle of Skye**
Geology
page 8

**Key**
Tees-Exe line:
– – – the divide between Highland and Lowland Britain
Land over 200 metres (major areas)

**Lake District**
Highland glaciation
pages 84–87

**Mappleton**
Coastal erosion
pages 71–73

**Afon Nant Peris, Snowdonia**
Fieldwork
page 47–51

**Midlands**
Floods 1998
pages 66–67

**Southern England**
Great Storm 1987
page 27

**The Severn Estuary**
Enquiry pages 76–77

**Selsey**
Tornado 1998
page 27

0          200 km

**The Mendips**
Slope studies
pages 92–93

**Studland, Dorset**
Sand dunes
pages 38–39

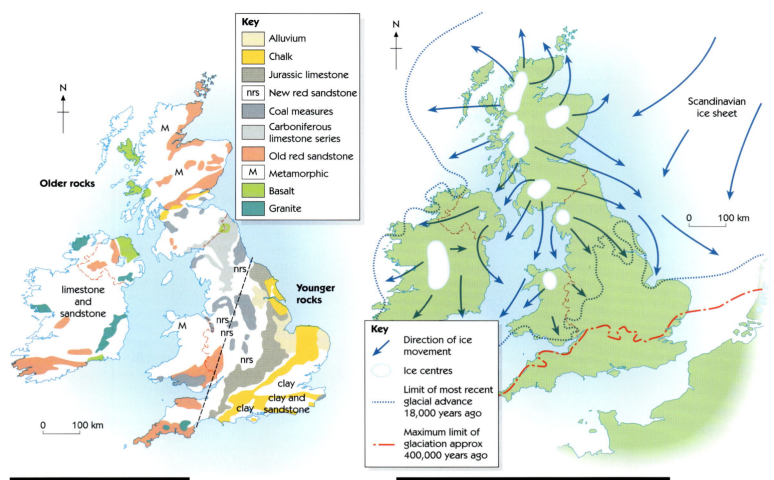

**Figure B: Geological map of the UK**

Key
- Alluvium
- Chalk
- Jurassic limestone
- nrs New red sandstone
- Coal measures
- Carboniferous limestone series
- Old red sandstone
- M Metamorphic
- Basalt
- Granite

Older rocks

limestone and sandstone

Younger rocks

nrs
nrs
nrs
M
nrs

clay
clay and sandstone
clay

0    100 km

**Figure C: The Ice Age and its effects on the British Isles**

Scandinavian ice sheet

0    100 km

Key
- Direction of ice movement
- Ice centres
- Limit of most recent glacial advance 18,000 years ago
- Maximum limit of glaciation approx 400,000 years ago

Generally, higher land is to the north and lower land is to the south, but there are exceptions.

The physical background of the UK includes more than relief and rocks. The pattern of climate, both present and past, also influences the physical geography. Figure C shows the extent of glacial climates in the UK. The country's climate does seem to be changing and by 2030–2050 London may have a similar climate to present-day Paris. Temperatures in Scotland may well rise by 1.6°C during the same time.

The soils and natural vegetation of the country is determined by the rocks, relief and climate. Figure D shows the make-up of the physical basis of geography. Notice how each part of the **ecosystem** is interrelated – everything depends on and influences everything else. People have interfered with the physical basis of the UK and they continue to do so, as many of the case studies in this book show. There is little really natural landscape left in the country, but the graph, Figure E, shows that about 30% may be semi-natural.

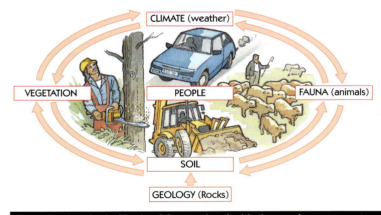

**Figure D: The physical basis of Geography: the biotic complex or ecosystem**

CLIMATE (weather)

VEGETATION          PEOPLE          FAUNA (animals)

SOIL

GEOLOGY (Rocks)

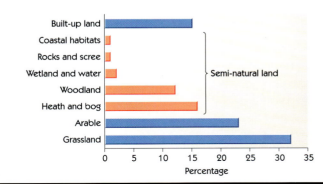

**Figure E: Only 30% of the country's land is semi-natural**

Built-up land
Coastal habitats
Rocks and scree
Wetland and water
Woodland
Heath and bog
Arable
Grassland

Semi-natural land

0    5    10    15    20    25    30    35
Percentage

## Do you know?

? The three rock classifications sedimentary, igneous and metamorphic

? The climate of the British Isles is called **mid-latitude temperate oceanic** The natural vegetation is a mid-latitude mixed forest, and is a mixture of broadleaf and coniferous trees.

# 2 Plate tectonics

**Do you know?**

? There are over a dozen crustal plates covering the Earth's surface.

? The plates move slowly across our planet's surface.

? Movements at the margins of the plates help to explain earthquakes and volcanic eruptions.

**Key ideas**

● The Earth's crust is divided into a number of separate sections called plates.
● There are three types of plate margin: constructive, destructive and conservative.
● Earthquakes and volcanoes are concentrated along the margins of the plates.

The ground beneath our feet may seem stable and secure. We say 'as safe as houses' and 'as old as the hills' – but our houses are not safe, and the hills, in terms of the age of the Earth, are quite young. This is because we live on a restless planet. The ground beneath our feet is the top of a thin crust of solid material above a fiery layer of molten rock.

The Earth's crust is divided into over a dozen separate sections called plates (Figure A).

## Types of plate margin

The plates of the Earth's crust are in constant motion, moving very slowly across the planet – at between 1 and 15 cm per year. They are driven by heat from the Earth's interior. Most of the world's volcanoes and earthquakes are found at their edges, where two plates meet.
There are three types of plate margin: **constructive, destructive and conservative**.

## Constructive plate margins

Constructive plate margins are also called **divergent margins** because they occur where two plates are moving apart (Figure B). Hot magma rises from within the Earth's interior and spills out onto the seafloor as **basalt**, making **oceanic ridges** totalling 70,000 km in length. The Mid-Atlantic Ridge is an example. This is a range of volcanoes running in a narrow line beneath the middle of the Atlantic Ocean. In places, these volcanoes are so high that they rise above the ocean surface to form islands such as Iceland and the Azores. As the plates move apart, the ocean widens in a process known as **seafloor spreading**. Africa and America were joined together 100 million years ago: they have spread 3000 km apart, a rate of about 3 cm per year.

Constructive margins on land form rift valleys such as the Thingvellir Graben in Iceland and the Adar Depression in East Africa. The East African Rift Valley is thought by some geologists to represent a failed plate margin or evidence of the early stages of the break-up of Africa The Red Sea is thought to be an early sign of the birth of a new ocean.

**Figure A: World map with crustal tectonic plates**

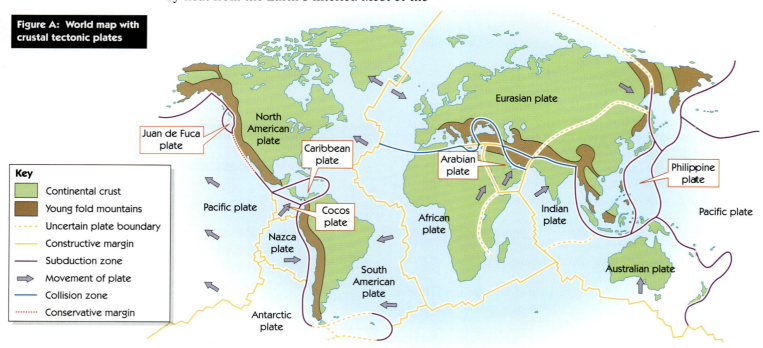

**Key**
- Continental crust
- Young fold mountains
- Uncertain plate boundary
- Constructive margin
- Subduction zone
- Movement of plate
- Collision zone
- Conservative margin

## Destructive plate margins

Destructive plate margins are also called **convergent margins** because they occur where two plates are moving together. There are two types of destructive margin: subduction zones and collision zones. The sort of crust involved determines which type of destructive margin will be formed.

Subduction zones occur at the deep ocean trenches where the sea floor is pulled down as one plate slowly passes under another (Figure C). The subduction zone is marked by earthquakes whose foci (sources) form a sloping line going deep under ground. As the oceanic crust plunges into the mantle of molten rock it melts and the less dense material rises to the surface where it erupts to form explosive volcanoes and violent earthquakes. A chain of volcanic islands called an **island arc** is formed – such as the Aleutian Islands off Alaska.

Continental crust is less dense than oceanic crust. When plates containing continental crust converge at collision zones, mountains are thrust upwards (Figure D). The force of such collisions radiates across thousands of kilometres, spawning earthquakes at great distances from the margin. For example, the Indian plate is colliding with the Eurasian plate, creating the Himalayas; earthquakes from this tremendous collision occur regularly as far away as Iran and central China.

## Conservative plate margins

Conservative plate margins are also called **transform margins** because they occur at transform faults in the crust. Here two plates simply move past each other (Figure E). No crust is created or destroyed at conservative margins and there is no volcanic activity. However, the movement is not continuous but occurs in a series of jerks which cause earthquakes whose foci are close to the surface. An example of a conservative margin is the San Andreas fault system in California where the Pacific plate is moving past the American plate.

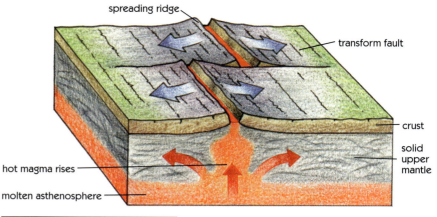

Figure B: Constructive plate margin

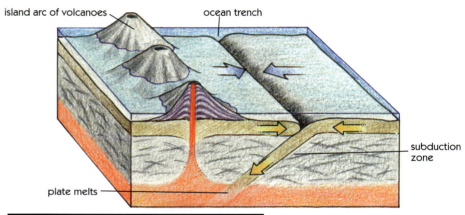

Figure C: Destructive plate margin: subduction zone

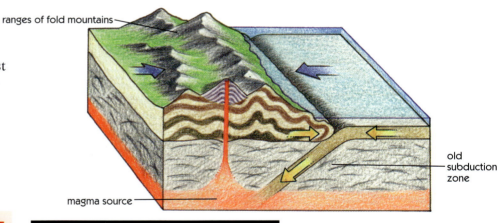

Figure D: Destructive plate margin: collision zone

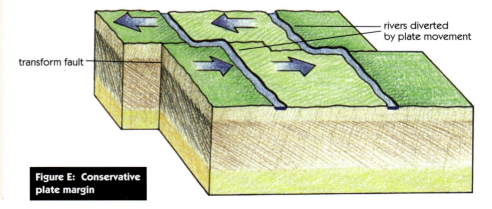

Figure E: Conservative plate margin

### ▼ Questions

1 a What is a tectonic plate?
  b How fast do the plates travel across the Earth's surface?
  c What events occur at plate margins?
2 Name the three types of plate margin.
3 What is the Mid-Atlantic Ridge and how is it formed?
4 a With what type of plate margin are ocean trenches associated?
  b Why are some destructive plate margins marked by subduction zones and others by collision zones?

# Earthquakes

Afghanistan is on a major plate boundary. To the south the Indian plate moves relentlessly northwards while to the north the vast Eurasian plate moves south-eastwards. The resulting collision has been under way for fifty million years and has created the world's highest mountain range, the Himalayas. It is not surprising that Afghanistan suffers many earthquakes, from frequent small tremors to occasional large earthquakes. 1998 was an exceptional year, however, because the unfortunate country, already devastated by civil war, was hit by two huge earthquakes in the same region, the foothills of the Hindu Kush mountains.

developing countries kills many more people than one of similar magnitude in a developed country (see the Italian example opposite). Why is this?

- The houses were not designed to withstand earthquakes. They were mainly mud brick with shallow foundations. The villages were built on steep, unstable slopes. They were swept away by mudslides.
- The area is so remote that it was two days before word of the first earthquake reached the outside world. No modern telecommunications were available. It took two more days for the first relief workers to reach the scene: the local roads had cracked open and been destroyed by landslides.
- It was very difficult to get relief equipment through. There were no local airstrips. Only helicopters could reach the disaster area; the only helicopters available in the early days were three small United Nations machines which could not carry much. As a result rescuers were forced to work with their bare hands as they tried to dig out the injured from the rubble.
- It was not possible to evacuate people quickly from the area: many died from injuries which would not have killed them if they had received basic medical treatment quickly. An aftershock three days after the February earthquake killed over 250 more people.
- International relief efforts were mounted after both earthquakes. Governments flew relief supplies into Faizabad airfield. Medical staff and other relief workers flew in from the USA, the UK, France, Turkey, Iran, Tajikistan, Russia and several other countries. Unfortunately the relief effort was too late to provide much help to victims of the earthquakes; efforts were concentrated on helping the survivors who were not injured.

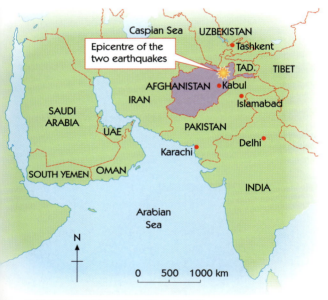

**Figure A: The location of the 1998 Afghanistan earthquakes**

## 4 February 1998

- 2400 people were killed when an earthquake measuring 6.1 on the Richter Scale hit the city of Rustaq 250 km north of the capital Kabul.
- 15,000 homes were destroyed, mainly by landslides triggered by the quake.

## 30 May 1998

- Over 5000 were killed by an even more powerful earthquake measuring 6.9 on the Richter Scale which struck the same area.
- Over thirty villages were completely destroyed and another seventy suffered serious damage.
  Afghanistan, like many poorer developing countries, was ill-equipped to face such natural disasters. It often happens that an earthquake in

**Figure B: Damage caused by the 1998 Afghanistan earthquakes**

## ▼ Questions

1.  a  Where were the Afghanistan earthquakes?
    b  What caused the earthquakes?
    c  What damage did the earthquakes cause?
2.  Why did so many people die in the Afghanistan earthquakes?

Italy experiences many small earth tremors and occasional large earthquakes. They occur as the African plate moves northwards towards the European plate. On 26 September 1997 the hill town of Assisi in central Italy was hit by a series of earthquakes (see Figure C). The largest shock was 6.0 on the Richter scale.

- 10 people were killed
- hundreds were left homeless
- damage was done to the thirteenth-century Basilica of St. Francis of Assisi. This is built halfway up a mountain making it vulnerable to the effects of earthquakes
- old frescoes (wall paintings) were damaged and the Mayor said 'We have lost part of the world's artistic wealth.'
- 70% of the town's buildings were evacuated for fear of people's safety
- an emergency unit was set up in the main square
- a tent city was set up for evacuees on the edge of the town
- art restorers and technical experts were on the scene quickly to assess damage
- the oldest buildings in Assisi and in other surrounding towns were damaged
- international help from UNESCO was offered to salvage damaged artworks.

Although earthquakes in Italy can be devastating, the damage is rarely as severe as that suffered in the poorer, less economically developing countries. This is not because the earthquakes are more powerful in poorer countries such as Afghanistan. It is more to do with the ability of richer countries to cope with severe quakes. Figure D shows why damage might be less in a country like Italy than in one such as Afghanistan.

Figure C: Location of Assisi

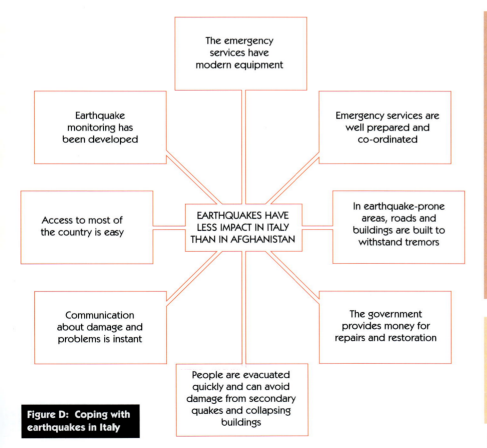

Figure D: Coping with earthquakes in Italy

## ▼ Questions

**1** Where was the Italian earthquake? Why did it occur? What damage did it cause?

**2** In what ways did Assisi receive help?

**3** In what ways did the effects of the 1997 Italian earthquake differ from those of the 1998 Afghanistan earthquake? Answer using columns set out like this:

|  | ITALY | AFGHANISTAN |
|---|---|---|
| Level of Wealth |  |  |
| Injuries |  |  |
| Costs |  |  |
| Response Time |  |  |
| International Help |  |  |

### Review

Afghanistan and Italy have both suffered earthquakes. The extent of damage was greater in Afghanistan because of the difference in the level of development in the two countries.

# Volcanoes

## Main activity

Interpreting information, and writing an explanation for local people.

In this unit on volcanoes the case study of the Montserrat volcano will be used to begin to answer several key questions.
● What caused it?
● Why did it occur?
● What impact did it have on the landscape?
● What effect did it have on the people?
● What was done to help the people?
● Can we predict the future for Montserrat?

People do not think that they will be affected

The volcano has not yet erupted during their lifetime

They have their homes and their farms in the area

The soils are fertile on the old volcanic ash and broken down lava

There is a tourist industry based on the sights of the local volcano

The local people cannot afford to move

The locals have never considered moving

People would not know where to move to

**Figure A: Why people live near volcanoes**

It is estimated that 500 million people live near enough to a volcano to be affected by a very severe eruption. So why do so many people live in possible danger? Figure A gives some answers. These were also the reasons why people lived in Montserrat, a small island in the Caribbean island arc named the Lesser Antilles.

The Soufrière Hills in the south of the island of Montserrat erupted 400 years ago. Then on 18 July 1995 there was an unexpected eruption. There was a violent explosion as magma met groundwater. Steam explosions were accompanied with eruptions of **tephra** which is a general name for volcanic ash particles. Alongside the eruptions were earthquakes. The timetable in Figure B shows that the 'event' was to continue for years, not just for days, which is usual for a volcanic eruption.

Stories of human suffering filled the world's newspapers. Daily bulletins could be followed on the Internet. One such story was about Maria Skerritt. Like hundreds of other Montserrat residents, Maria Skerritt abandoned her home and small vegetable farm in the central part of the Caribbean island after the Soufrière Hills spewed out hot ash and rocks in June 1997. The reasons for the volcanic eruption are shown on the map (Figure C) and the diagram of the plate margin (Figure D). It happened in 1995 because the movement of the plates had built up so much pressure that it could no longer be contained. The magma rising beneath the Soufrière Hills is re-melted material that was once part of the North American and South American plates which has sunk beneath the Caribbean plate. As it came to the surface it formed a steep sided **dome**. The material forming it is extremely sticky and viscous (meaning it does not flow very freely). By contrast, in Hawaii the lava from the volcanoes is non-viscous and forms gently sloping volcanic sides.

## Helping the people

- Monitoring devices were set up
- The British government gave £41 million relief
- A 'safe zone' was established and people were moved there (Figure E)
- Countries sent medical assistance. In August 1997 Cuba sent doctors and nurses
- Promises of new housing were replaced by evacuation packages
- Help was given to improve infrastructure in the north
- Families moved to Antigua, USA and Britain

Local people were obviously very traumatised and did not accept early evacuation. Many thought the evacuation warnings came too soon. The amount of money offered by the British government was seen as too small.

People were angry at the lack of facilities, such as lavatories and fridges, in the north.

**18 July 1995**
Steam and tephra eruptions continuing for a few days

**Mid-August**
6000 people had been evacuated to the north of the island

**21 August**
An eruption column of 2,100 metres; tephra fell on Plymouth, see Figure E, and it was dark for 30 minutes

**August to December**
A new lava dome grew, see Figure E.

**December**
Plymouth evacuated as a precaution

**December to April 1996**
Lava eruptions occurred leaving large spines rising out of cracks in the new dome. These often collapsed. Residents allowed back to Plymouth and the south

**April 1996**
Tephra columns up to 12,000 metres. Pyroclastic flows developed. These are flows of superheated gas, volcanic rock and minerals flowing like a fluid

**April 1997**
Pyroclastic flows started again and tephra fell on the north and the south

**June 1997**
The volcano became very dangerous

**25 June**
Part of the dome collapsed. 9000 metre tephra plume. Pyroclastic flow nearly reached the airport. 7 villages and 175 homes were destroyed. 19 people died

**July**
See Figure A

**4–8 August**
Plymouth destroyed. 80% buildings destroyed including the seat of government, the customs office and public market. 12,000 metre plumes. Ash fell on nearby Nevis to the north, see Figure D. Some of the pyroclastic flows moved at 100mph and reached 1000°C. See the devastation in Plymouth on the Introduction page 5.

**September**
Two-thirds of Montserrat had been destroyed. 7000 of the 11,000 population had left the island

**26 December 1997**
The largest magnitude and the most intense eruption so far

**19 March 1998**
4 earthquakes and 9 rockfalls

**March to July**
Volcano quiet (dormant)

**3 July**
Molten rock and mud poured out. 10,000 metres of ash. Warnings to aircraft. No injuries to those living in the northern 'safe zone'

The Montserrat volcano had not finished. People who had pinned their hopes on returning were devastated

**September 1998**
Volcano active again. Smell of sulphur as far away as Nevis

Figure B: Timetable of the the Montserrat volcano

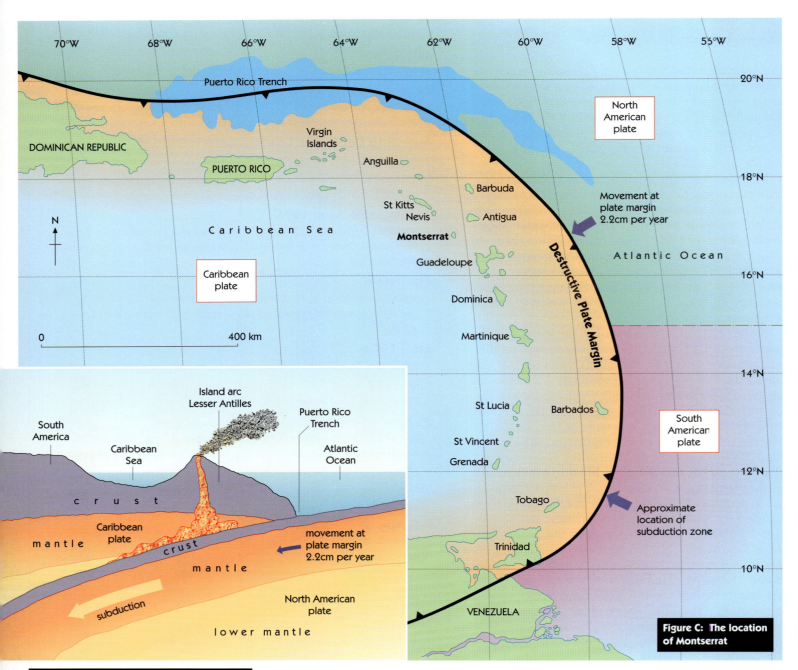

Figure C: The location of Montserrat

Figure D: The plate margin under Montserrat

## Monitoring the volcano

Scientists from the United States Geological Survey, the British Geological Survey and the University of the West Indies have been monitoring the Montserrat volcano since 1995.

1 Using seismic instruments to measure earth tremors. These occur as magma is rising up the volcano before it erupts. There is however often no warning of a tremor.
2 Using equipment which measures the levels of gases in the atmosphere and changes in the acidity of rainwater – see Figure F.
3 Measuring the tilt or deformation of the ground very accurately. This is done by using tiltometers. A helicopter paid for by the British government makes daily flights which record the changing dome and pyroclastic flows on the mountain.

These monitoring methods should build up a picture of the pattern of events so that prediction will be possible in the future. It is not easy – the next eruption may be 400 years from now! At present it is not even possible to predict that the eruptions have finished. The people cannot be advised to return. Many are beginning new lives in a foreign country.

## Developing a housing programme for Montserrat

Rick Groves is a senior lecturer at Birmingham University and he researched the problems of housing on Montserrat. He commented on the difficulties that prevented a quick house-building programme.

● Because the volcano was still active no new insurance cover was available, making it difficult to obtain loans or mortgages for house building.

Figure E: Map of Montserrat showing zones

NORTHERN 'SAFE ZONE'

• St. Johns

CARIBBEAN SEA

CENTRAL AREA
Various definitions over the hazard time

• Salem
■ Montserrat Volcano Observatory

Airport

Spanish • Point

SOUTHERN 'NO ACCESS' ZONE
▲ New Dome

Plymouth (capital) •

Soufrière Hills

N

0        2 km

• St. Patrick's

Figure F: A BGS seismologist measures gas from the volcano

● There was a lot of communal land ownership in Montserrat and it was not clear who owned much of the other land.
● It was the north of the island that was safe but it was this part that had the poorest roads, drains and power sources (**infrastructure**).

The main aims of the future house-building programme will be:

● to build houses for people with priority needs
● to provide a programme of community care
● to manage and maintain the new houses.

Apart from its practical effects, the house-building programme will have a deeper significance for the population of Montserrat, demonstrating as it does a commitment to the island's future.

Figure G shows the type of land in the northern safe zone where new housing has been built and where many people of Montserrat will probably settle in the future. Figure H shows two new types of housing that were built quickly after the crisis. In the foreground is a temporary wooden building designed to house people collectively. In the background are the first system-built houses which were put up as quickly as possible. Both types of housing were expensive to build but neither has been very successful.

Figure H: Safe housing in the north

Figure G: North side of Montserrat

The Montserrat volcano ruined the island and the lives of many people. There were 19 deaths. Assistance saved many lives and helped others to begin again. The volcano was a long-term event which may continue. Monitoring was well established but prediction has not been possible.

## ▼ Questions

1 Look back to the six questions asked at the beginning of the unit (page 16). Write brief answers to these questions in note form only.

2 Why were people not always pleased with the assistance and help offered?

3 What sort of monitoring was set up? Divide your answer into seismic, environmental and deformation methods.

4 Write an article of no more than 200 words about the Montserrat volcano. Choose a title such as 'Caribbean paradise island destroyed'.

5 Write a simple account of the reasons for the volcano that can be presented to local people. They are saying that they want to return to the island as soon as the eruptions stop. You must put the case to them that it is still not safe to return. ➡

6 Study the housing photograph, Figure H. Why do you think local people did not find these two types of home very easy to settle in to?

# Weather

## Key ideas

● Britain's changeable weather is mainly due to the frequent passage of depressions across the country.
● Anticyclones bring long periods of settled weather.

## Do you know?

❓ Atmospheric pressure is caused by the weight of the atmosphere upon the Earth's surface.
❓ Warm air can hold more water vapour than cold air.

The everyday changes in the atmosphere over an area is called its weather. In Britain the weather conditions often change very quickly. We can measure and record the weather using a number of scientific instruments.

## ▼ Question

**1** Use a dictionary to help you link the weather process with the correct instrument for recording it.

| | |
|---|---|
| Temperature: | *barometer* |
| Wind speed: | *thermometer* |
| Atmospheric pressure: | *hygrometer* |
| Humidity: | *anemometer* |

## Pressure

The air above us is pressing down upon the surface of the Earth. This is atmospheric pressure, and it changes from place to place and from time to time.

● As air rises, this reduces the weight of air at the Earth's surface and so creates low pressure.
● As air descends, this increases the weight of the air at the Earth's surface and creates high pressure.

Pressure differences are caused by temperature differences: hot air rises, cold air descends. Winds are caused by pressure differences between one place and another; winds blow from high pressure areas towards areas of low pressure. Atmospheric pressure is measured in millibars. The average atmospheric pressure at sea level is 1013 millibars.

## Temperature

Why do temperatures vary from place to place? Among the most important reasons are:

● height above sea level. Temperature drops by about 1°C for every 100 metres above sea level. Hilly and mountainous areas are thus much cooler than surrounding lowlands.

**Figure A: How the angle of the sun's rays causes temperatures to vary**

When the sun's rays hit the surface vertically a small area is intensely heated

The same amount of heat is spread over a much larger area when the sun's rays hit the surface at an angle. Much less heat is available per unit area.

● the angle of the sun's rays. An area receives more heat when the sun shines from a high angle overhead and less heat when the sun's rays reach it at a lower angle (Figure A). This causes differences between winter and summer temperatures. It also means that places further from the tropics have lower temperatures.
● distance from the sea. Water heats up more slowly than the land, but is able to store the heat much longer. This means that in summer the sea is usually cooler than the land, but in winter the sea is warmer than the land. Places near the sea have cooler summers and milder winters (a maritime climate) than places further inland (a continental climate).

## Water in the atmosphere

Water is present in the atmosphere in three forms:
● as a liquid
● as a solid – ice or snow
● as an invisible gas called water vapour.

All air contains water vapour, even in the driest desert. A parcel of air can only hold a certain amount of water vapour. The amount of water vapour which the air can hold depends upon its temperature: colder air holds less water vapour than warm air. Warm air rises. As air rises, it expands and cools. As the air cools some of the water vapour contained in it condenses into water droplets and these water droplets form clouds. The droplets grow larger and heavier by joining together. If the tiny cloud droplets become too heavy to remain in the cloud, it will rain.

## ▼ Questions

2 a What causes differences in atmospheric pressure?
  b What causes winds?
3 How is temperature affected by:
  a height above sea level
  b the angle of the sun's rays
  c the distribution of land and sea?
4 a What is water vapour?
  b How does the temperature of the air affect the amount of water vapour which the air can hold?

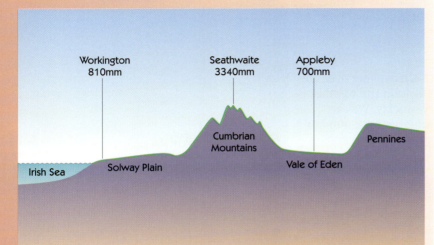

**Figure B: A cross-section through the Lake District showing annual rainfall**

5 Study Figure B.
  a What is the annual rainfall at:
     (i) Workington (ii) Seathwaite (iii) Appleby?
  b Why does this pattern of rainfall happen?
  c Appleby is in a **rain shadow** area. What does this mean?
6 Use your atlas map of the climate of the British Isles.
  a Find the map showing January temperatures. Name areas of the British Isles which have average January temperatures:
     (i) over 6°C (ii) below 4°C.
  b Find the map showing July temperatures. Name areas of the British Isles which have average July temperatures:
     (i) over 16°C (ii) below 13°C.
  c What factors appear to affect the pattern of January and July temperatures across the British Isles?
  d Use the map of annual rainfall over the British Isles.
     (i) Name five areas which have over 2000 mm annual rainfall.
     (ii) Which areas have less than 750 mm of annual rainfall?
     (iii) What factors appear to affect the pattern of annual rainfall over the British Isles?

## CASE STUDY: Depressions and anticyclones over Europe

The low pressure system, or depression, is the most important weather system which affects Britain and North West Europe. In Britain, and in the northern hemisphere as a whole, winds blow inwards towards the centre of a depression in an anticlockwise direction. Depressions bring cloud and rain. Figure C shows a depression which passed across western Europe on 9–10 February 1996. It is a huge feature, stretching hundreds of kilometres across from Iceland to Spain. This depression formed out over the Atlantic Ocean. Depressions develop where warm air meets cold air. Warm air rises over the colder air to form a **warm front**. Heavier cold air moves in from behind to form a **cold front**. It is the changes in position of the two types of air which cause the wind, cloud and rain associated with a depression, as Figure D explains.

The three images in Figure C show very clearly the cold front stretching from Ireland across the Bay of Biscay, moving inland across France. By 0200 on 10 February the cold front runs through the North Sea, across the Netherlands and eastern France. The warm front is also visible, but is not such a clear feature.

The depression moves eastwards taking its wind and rain with it. As the depression passes overhead, a place will have a sequence of weather, mostly wet and windy. The warm air (the warm sector of the depression) gradually rises higher until it loses contact with the ground. The warm and cold fronts join to form an **occluded front**, and the depression quickly fades and dies.

0351 9 February

1346 9 February

0200 10 February

**Figure C: Three satellite images showing a depression**

Approx scale across France

0 — 400 km

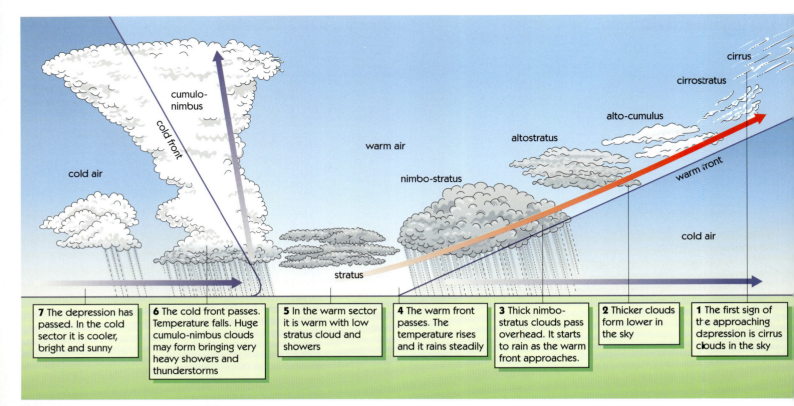

| 7 The depression has passed. In the cold sector it is cooler, bright and sunny | 6 The cold front passes. Temperature falls. Huge cumulo-nimbus clouds may form bringing very heavy showers and thunderstorms | 5 In the warm sector it is warm with low stratus cloud and showers | 4 The warm front passes. The temperature rises and it rains steadily | 3 Thick nimbo-stratus clouds pass overhead. It starts to rain as the warm front approaches. | 2 Thicker clouds form lower in the sky | 1 The first sign of the approaching depression is cirrus clouds in the sky |

**Figure D: A cross section through a depression**

## Anticyclones

An anticyclone is in some ways the opposite of a depression: it is a high pressure area, with light winds which blow outwards from the centre in a clockwise direction. There is little or no rain associated with an anticyclone. The weather in an anticyclone varies according to the season.

● In the summer anticyclones bring hot, sunny, dry weather. If an anticyclone remains over Britain during the summer for more than a few days we speak of a 'heat wave'.

● In winter, however, the weather can be much less pleasant. Since there are few clouds with an anticyclone, the Earth's heat escapes quickly at night. This causes frost on the ground and it becomes very cold. Winter anticyclones often have fog. The cold ground causes water vapour in the lower air to condense into droplets. These hang in the air until the heat of the sun raises the temperature. Because the winter sun is weak, the fog may remain all day.

### ▼ Questions

1 What do you understand by the following terms?
a   a warm front
b   a cold front
c   a depression
d   an anticyclone.

2 Study Figure C. Using your atlas, calculate the speed of the cold front moving across northern France. Is it:
a   20 kph
b   50 kph
c   90 kph?

3 Using Figure C, prepare a weather forecast to be broadcast at 0200 on the morning of 9 February to predict the weather for London over the next 24 hours. To accompany your forecast script you should include three weather maps with the usual TV symbols for sunshine, cloud, rain and wind. If possible, use a video camera to record your weather forecast.

4 How and why does the weather in an anticyclone vary between winter and summer?

## Key ideas

● Storms are violent, short-lived events. They may have short- and long-term impacts upon people's lives and activities.
● Tornadoes may develop where there are strong updrafts of air along cold fronts.
● The USA suffers from tornadoes especially in the mid-West in springtime.
● The UK faces a variety of storm hazards including tornadoes and storm force winds.

## Do you know?

? Tornadoes are much more common in the UK than most people think. An average of 33 are reported each year.
? In 1987 Southern England was hit by the worst storm for over 250 years.
? El Niño is a periodic warming of the Pacific Ocean waters off Peru. It results in a temporary upset of the world's weather and may be responsible for weather hazards occurring in places that do not usually experience them. 1998 was an El Niño year and may have been responsible for the Florida tornadoes.

## Main activity

Interpreting information and report writing.

A tornado is a violent whirling wind which accompanies a funnel-shaped cloud. It forms when there is a strong updraft of air accompanying a cold front and a thunderstorm with a large **cumulo-nimbus** cloud. Figure A shows the five stages of the formation of a tornado. It is usually easy to see the funnel because of the dust which is sucked up. The speed of the funnel winds may be more than 480 kph (more than 300 mph). A scale for measuring the intensity of tornadoes can be seen on page 27. In the USA a similar scale called the **Fujita Scale** classifies tornadoes from F0 to F6.

There is always the risk of damage from storms and tornadoes. The force of the winds damages vegetation and property. Buildings and other structures may explode as the pressure is suddenly extremely reduced as the tornado passes over. This mainly happens when buildings are not ventilated to adjust to the

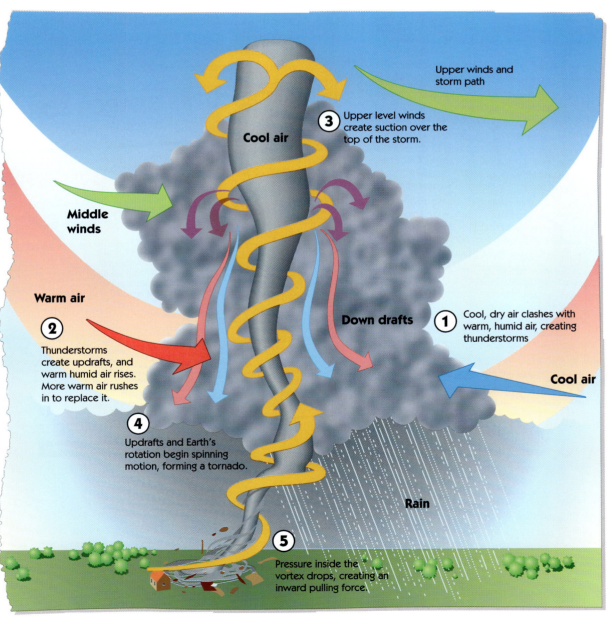

Upper winds and storm path

**Cool air**

③ Upper level winds create suction over the top of the storm.

**Middle winds**

**Warm air**

② Thunderstorms create updrafts, and warm humid air rises. More warm air rushes in to replace it.

④ Updrafts and Earth's rotation begin spinning motion, forming a tornado.

**Down drafts**

① Cool, dry air clashes with warm, humid air, creating thunderstorms

**Cool air**

**Rain**

⑤ Pressure inside the vortex drops, creating an inward pulling force.

**Figure A: How a tornado forms**

Tornadoes are most common in the mid-West of the USA including Texas. They start in springtime and move north during the summer months. In 1998 the tornado season started earlier and further south. It is thought that El Niño was to blame. The high level **jet stream** was further south during the winter of 1998 and cool dry air was meeting warm humid air over Florida – see Figure A. Figure C shows that early 1998 was a very bad period for tornadoes in the USA.

|  | 1995 | 1996 | 1997 | 1998 | Average |
|---|---|---|---|---|---|
| January | 2 | 1 | 2 | – | 2 |
| February | 6 | 1 | 1 | 42 | 3 |
| March | 1 | 6 | 28 | 16 | 11 |
| April | – | 12 | 1 | 53 | 5 |
| May | 16 | 1 | 29 | 10 | 15 |
| **Total** | **25** | **21** | **61** | **121** | **36** |

**Figure C:  Number of tornado deaths in the USA 1995–1998**

## EL NINO BREWS MIDWEST-STYE WHIRLING WINDS

By Mike Oliver and Seth Borenstein

Texas-sized twisters don't belong here.

Florida is known for coast-crashing hurricanes, blinding tropical rainstorms and relentless summer sunshine. But relative to other parts of the country, Florida tornadoes are usually wimpy, hopscotching whirlwinds whose havoc-wreaking is contained to mobile-home porches and the occasional toppled tree.

'What you saw was Midwest tornadoes in Florida,' said Craig Fugate, Florida's chief of disaster preparedness and response.

Before this storm, the last recorded death resulting from a tornado in any of the affected counties – Orange, Osceola, Seminole and Volusia – was in 1959, according to the National Weather Service.

So the utter devastation Monday in the aftermath of the state's worst tornado outbreak seemed more akin to Texas, Kansas or some other 'tornado alley' state.

El Nino, a periodic warming of Pacific Ocean waters off the coast of Peru, pushes southward the strong river of winds called the jet stream, which circles the globe at jet-plane altitudes. With the jet stream lying across Central Florida, the intense storms whip into Florida from the Gulf with frightening speed, experts said.

**Figure B:  From The Orlando Sentinel, 24 February 1998**

In February 1998 the worst ever tornadoes hit central Florida. The newspaper articles (Figure B and D) from the *Orlando Sentinel*, Tuesday 24 February gives some of the horrific details. The paper was brought home by a girl who was on holiday in Orlando at the time.

The Florida tornadoes killed more people than the infamously destructive Hurricane Andrew in August 1992. Why was this, when weather alerts and warnings have been so well developed in the USA? In Florida itself there is a multi-dollar, high-tech weather radar and warning system. However, people have to be awake to listen to it. The storms hit in the middle of the night when people were asleep.

## RESCUERS SEARCH FOR MORE VICTIMS
### 10 remain missing – 250 injured

By Mike Oliver

A devastating sweep of midnight tornadoes left so much destruction across Central Florida on Monday that rescue workers continued to struggle early today to find or identify all the dead.

So far, at least 36 are confirmed dead, 250 injured and about 10 still missing from the deadliest twisters in state history.

Incredible stories emerged in the aftermath: An 18-month-old toddler was found alive in a tree. A trucker ushered his fiancee into a closet just before he was blown from his home to his death. A man, his 79-year-old mother and 98-year-old grandmother huddled in prayer as the storm raged outside, but their home was spared.

Makeshift morgues were set up at hard-hit areas, and dozens of the suddenly homeless trickled into shelters.

As victims surveyed damage to hundreds of homes in Osceola, Orange, Seminole and Volusia counties, Gov. Lawton Chiles declared Central Florida a disaster area and President Clinton made plans to visit Wednesday. The cost of the damage was far too great to estimate Monday.

'This was one for the century, one for the ages, you could say,' said Jim Lushine, a Miami meteorologist with the National Weather Service.

The previous Florida record for a tornado was 17 people killed directly by Hurricane Andrew in 1992. The death toll was so high Monday that the Orange County Medical Examiner's Office ran out of refrigerated storage space and was forced to borrow body bags from the Greater Orlando Aviation Authority, forensic co-ordinator Carol Gross said.

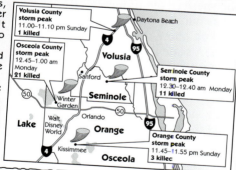

**Figure D:  From The Orlando Sentinel**

## ▼ Questions

*Either*

1 What is a tornado?
2 Why do they occur?
3 Why did they occur in Florida in February 1998?
4 What were the effects of the 1998 tornadoes?

*Or*

5 Write a report of the tornadoes that hit Florida. You should use the following structure for your report (a writing frame). What was it? Where was it? When was it? Why did it happen? Who was affected? What were the effects and problems?

### Review

The Florida tornadoes were the worst on record and were a result of the shift in the weather patterns because of the El Niño phenomenon.

On Wednesday 7 January 1998 the West Sussex seaside resort of Selsey was hit by a tornado. Just before midnight a waterspout moved onto the beach, lifting shingle and hurling it outwards, smashing the windows of some bulldozers on the beach. The tornado swept across open fields and then hit the first house in its path, removing the gable end and exposing the rooms within to view. Someone inside was badly cut by flying glass.

The tornado then tracked right through Selsey's town centre. Winds of over 160 kph caused extensive damage and made a frightening howling sound. Over one thousand buildings were damaged; sheds, fences, caravans and greenhouses were destroyed; trees and bushes were uprooted. An observatory belonging to the astronomer Patrick Moore was smashed.

Many people were moved out to emergency accommodation. 150 firefighters and emergency service workers were called in to make buildings safe and clear the damage. Dozens of people had minor injuries and shock, and the total damage caused by the tornado exceeded £2 million. Tornadoes are not so rare as most people believe; Selsey was hit by a similar tornado twelve years earlier on 21 November 1986.

## ▼ Questions

1 List the effects of the Selsey tornado of 1998 under the following headings:
*damage to property; damage to the environment; casualties; costs*

2 Figure B shows the Tornado Intensity Scale devised by Dr Meaden, an English meteorologist, in 1972. Study the scale and estimate the intensity of the Selsey tornado of 1998.

| Intensity | Wind Speed (kph) | Effects |
|---|---|---|
| 0 | 60–86 | Light Tornado: loose leaf litter raised in spirals; twigs snapped; some loose roof slates removed |
| 1 | 87–115 | Mild Tornado: small plants and heavy litter raised; minor damage to sheds, hedges and trees. More slates dislodged |
| 2 | 116–149 | Moderate Tornado: light caravans blown over; garden sheds and greenhouses destroyed; much damage to tiled roofs and chimney stacks. Large branches snapped off large trees, small trees uprooted |
| 3 | 150–182 | Strong Tornado: light caravans destroyed; garages and outbuildings destroyed; severe loss of slates |
| 4 | 183–219 | Severe tornado: Entire roofs removed from some houses; some gable ends collapse; numerous trees uprooted or snapped off |
| 5 | 220–255 | Intense Tornado: cars raised; weak buildings may collapse completely |
| 6 | 256–298 | Moderately Devastating Tornado: lorries raised; strong buildings lose entire roofs and perhaps also a wall |
| 7 | 299–339 | Strongly Devastating Tornado: many walls collapse; locomotives thrown over |
| 8 | 340–384 | Severely Devastating Tornado: cars carried great distances; most houses damaged beyond repair; steel framed buildings buckled |
| 9 | 385–430 | Intensely Devastating Tornado: many steel-framed buildings badly damaged; trains hurled some distance; complete de-barking of any surviving tree trunks |
| 10 | 431–480 | Super Tornado: utter destruction; buildings smashed and their fragments hurled over great distances |

(note: 10 kph is roughly equivalent to 6 mph)

**Figure B: Tornado intensity scale (devised by Dr G Terence Meaden, 1972).**

**Figure A: Damage caused by the Selsey tornado of 1998**

## CASE STUDY: It happened in the UK: The Great Storm of 1987

On Friday 16 October 1987, Southern England was hit by the worst storm since 1703. Winds exceeded 175 kph on the south coast and reached 150 kph at London's Heathrow airport. The storm, incorrectly termed a hurricane, was actually an intense depression caused by the meeting of very warm air from Africa and the Gulf of Mexico and cold air from the northern Atlantic Ocean (see Figure C). After forming in the Bay of Biscay it crossed north-west France and gained energy as it passed over the comparatively warm waters of the English Channel. The temperatures recorded at 10 p.m. on the night of 15 October on either side of the warm front show the extent of the temperature contrast. It was 18°C in Kent but only 7°C in neighbouring Essex. A central pressure reading of 958 millibars and a very steep pressure gradient brought disaster in the early hours of 16 October.

● 19 people were killed.

● Many houses collapsed and others were severely damaged. A block of flats at Fareham in Hampshire collapsed. Chichester Cathedral suffered considerable damage when one of the pinnacles on the tower was blown off by the wind and fell through the roof; several stained glass windows were also blown in.

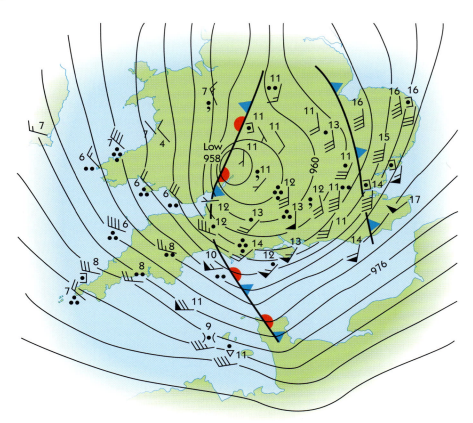

**Figure C: Synoptic chart for 16 October 1987, 0400 hours**

Figure D: Tree damage to a house and car at Addlestone, Surrey 16 October 1987

- A caravan park at Hayling Island was totally wrecked. At Selsey more than one hundred people were evacuated from a caravan site just before many of the caravans were wrecked.
- An estimated fifteen million trees were blown over, some blocking railways and roads and causing accidents. A third of the trees in London's Kew Gardens were ripped out and classic greenhouses worth millions of pounds collapsed. Twenty National Trust gardens and woodlands in the south of England were devastated. The trees suffered so badly because many had not yet lost their leaves and abnormally high rainfall in the preceeding fortnight loosened their roots.
- Power lines were brought down causing the worst power failures for forty years and blacking out much of southern England, affecting over 5 million homes.
- The London Fire Brigade received a record 6000 calls within 24 hours. Two firefighters died when a falling tree crushed the cab of their fire engine at Christchurch, Dorset.
- A 200m section of Shanklin Pier on the Isle of Wight was washed away.
- A cross channel ferry was blown aground near Folkestone. All the passengers were rescued by breeches-buoy; a bulk carrier capsized just outside Dover harbour and two of the crew drowned.
- Gatwick Airport was closed for several hours; a passenger on one of the last airliners to land before it closed spoke of passengers screaming as the aircraft bucked wildly before landing well down the runway.
- Almost 90% of the roads in Kent were impassable because of tree falls.
- The total cost in insurance claims was over £1000 million, of which £635 million was damage to domestic property. Over 1.2 million claims were received by insurance companies. This was the most expensive weather event the UK had ever suffered (but it was to be exceeded in the floods of April 1998, see page 66).

US meteorologists call such an event a 'weather bomb'. The speed with which the depression intensified caught British weather forecasters by surprise. They also predicted that the depression would track up the English Channel towards Belgium; instead it passed across southern England towards Denmark. There was a storm of criticism in the media about the lack of warning. Some of the damage could have been prevented if people had been warned, but it is possible that the warning may have caused a greater loss of life as more people may have exposed themselves to more risk by going out to watch the storm.

Following the 1987 storm the Meteorological Office introduced the 'severe weather warning' system and such warnings now feature regularly on TV weather forecasts.

### ▼ Questions

1. What was the cause of the Great Storm?
2. How did the storm affect human lives and activities?
3. a Why were British weather forecasters caught by surprise?
   b How do you think a more accurate weather forecast would have helped people?
4. In what ways were the Florida tornadoes different from the Selsey tornado?

### Review

Tornadoes are quite common in the UK, but the Selsey tornado of 1998 was very severe by UK standards.

# Climate

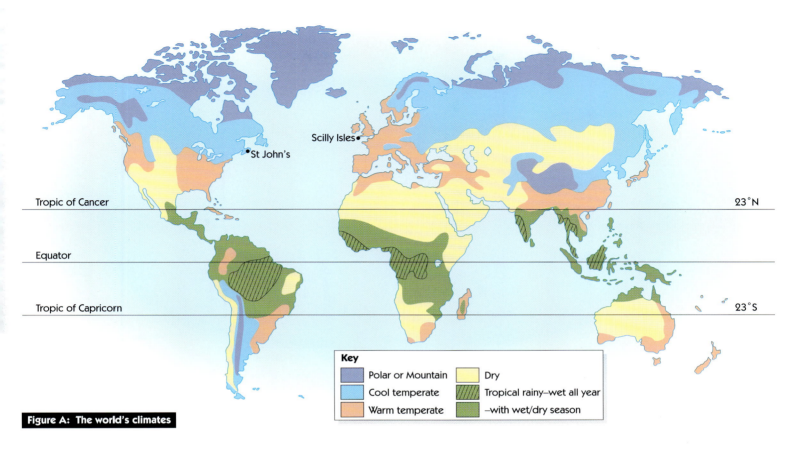

**Key**

| | |
|---|---|
| Polar or Mountain | Dry |
| Cool temperate | Tropical rainy—wet all year |
| Warm temperate | —with wet/dry season |

**Figure A: The world's climates**

## World distribution of climates

The earth has many different climates. Figure A shows a simple classification of the world's climates into five major types.

The hot deserts are the driest places in the world. However, many other places have unreliable rainfall. This is mainly because the prevailing winds are very dry; they move from the dry land out to the sea and usually pick up little moisture.

## The general atmospheric circulation

Figure B is a very simplified diagram of the atmospheric circulation. The general atmospheric circulation governs the prevailing winds, and is also closely related to the global distribution of climates.

Air in the equatorial regions is intensely heated. The hot air rises, causing low pressure at the surface of the land beneath. The rising air moves north and south towards the poles. The air which moves northwards is diverted to the right by the rotation of the Earth and it cools and sinks at about 30°N. Cool air then moves towards the Equator, completing a circulation known as a **convection cell**. The surface winds, blowing from the north-east, are called the **trade winds**.

There is another convection cell at the poles. The air is cooled, sinks and moves away from the poles. In between the two convection cells in the northern hemisphere is an area of winds blowing from the south-west. The circulation of the southern hemisphere mirrors that of the northern hemisphere.

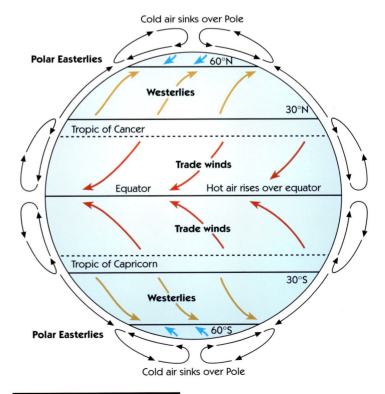

**Figure B: Atmospheric circulation**

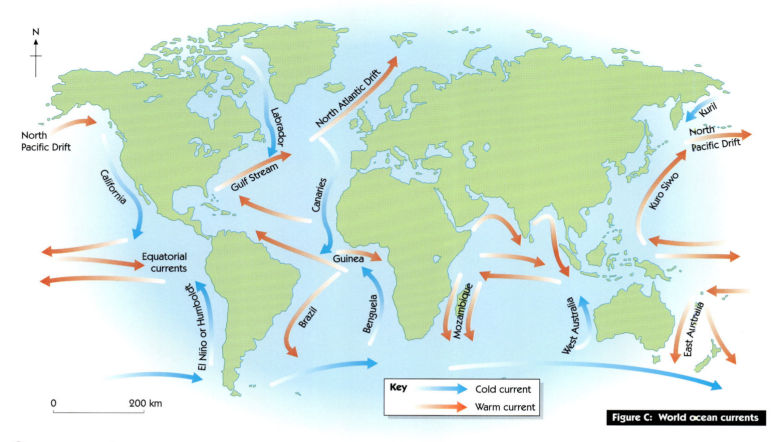

Figure C: World ocean currents

## Ocean currents

The winds (Figure B) are the major means of transferring heat from the equatorial and tropical regions to the temperate and polar latitudes. Ocean currents are another important means of heat transfer (Figure C). They are movements of surface water which can reach speeds of up to 3 kph. The major cause of ocean currents is the wind. There is a clear circulation pattern in the major oceans for each hemisphere: clockwise in the northern hemisphere and anti-clockwise in the south.

Figure D shows that ocean currents have a major influence on the climates of coastal areas. In winter, the temperature difference between the western and eastern coasts of the North Atlantic is most marked. Sea ice extends along the coast of Newfoundland, further south than the latitude of the English Channel. By contrast even the coast of northern Norway is ice free thanks to the warm current called the North Atlantic Drift and the prevailing south-westerly winds.

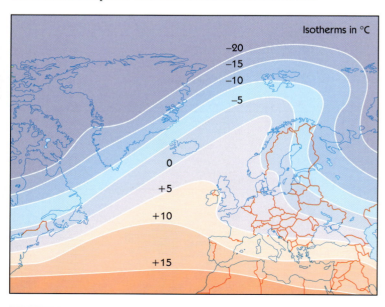

Figure D: Average temperatures in the North Atlantic in January

## ▼ Questions

**1** What is the main cause of low average rainfall totals?

**2** Study Figure B.
  a  Why is there low pressure at the Equator?
  b  Why is there high pressure at the poles?
  c  What is a convection cell?
  d  What causes the trade winds? How do you think these winds got their name?

**3** Draw a line graph to show the temperature figures for these two places on opposite sides of the Atlantic Ocean.

| Month | J | F | M | A | M | J | J | A | S | O | N | D |
|---|---|---|---|---|---|---|---|---|---|---|---|---|
| St John's, Canada (47° 37′ N) | −5 | −4 | −6 | −1 | 5 | 11 | 16 | 16 | 12 | 10 | 5 | −1 |
| Scilly Isles, UK (49° 55′ N) | 6 | 7 | 9 | 10 | 11 | 12 | 15 | 15 | 13 | 12 | 9 | 7 |

**RELATED PAGES** ▲RAINFORESTS: 42

## Main activity

Understanding and interpreting data

## Key ideas

● Where does the rainforest climate occur?
● What is a rainforest climate?
● Why is the climate hot with rain all the year?

In the western Amazon Basin area of Brazil (the state of Amazonas) there is a humid tropical climate with no dry season. This is often called an equatorial rainforest climate. The diurnal (daily) temperature range is usually greater than the average annual (yearly) temperature range. The evenings are hotter than early morning but January is as hot as July. The hot tropical sun heats up the land in the day and each night the land cools down. Each month the temperatures are about the same with a monthly average of 24–26°C.

The average annual rainfall in this area is shown in Figure A. Figure B shows the 'botanically dry' days in the Amazon year. São Gabriel do Rio Negro never has a month with rainfall below 150mm. Every month's

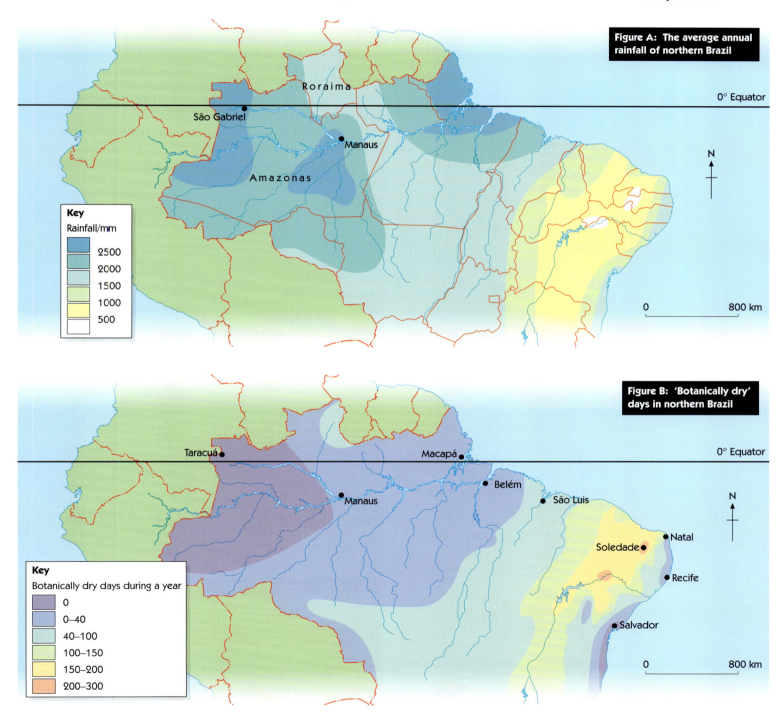

**Figure A:  The average annual rainfall of northern Brazil**

Key
Rainfall/mm
2500
2000
1500
1000
500

0° Equator

0       800 km

**Figure B:  'Botanically dry' days in northern Brazil**

Key
Botanically dry days during a year
0
0–40
40–100
100–150
150–200
200–300

0° Equator

0       800 km

average temperature is above 24°C. The climate graph for the São Gabriel meteorological station is shown in Figure C. Similar climatic areas are found in central Africa and the East Indies. Parts of the eastern Amazon rainforest do have a short drier season, but in most respects the climate is similar to that of the São Gabriel area.

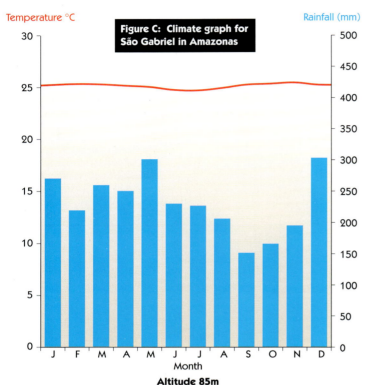

Figure C: Climate graph for São Gabriel in Amazonas

Temperature °C

Rainfall (mm)

Month

**Altitude 85m**
**Average annual temperature 25.2°C**
**Average range of temperature 1.6°C**
**Annual precipitation 2738mm**

It is worth considering a British climate station as a comparison. Penzance in the warm and moist south-west of England has no month with rainfall up to 150mm and no month with an average temperature above 16°C.

If you lived in a European climate area and visited the area around São Gabriel in Amazonas you would find the climate hot, muggy, sticky and uncomfortable. You would have to drink a lot to replace water lost by sweating. It would not be advisable to expose your body to the hot sun; loose full covering clothes would be best.

## Why so much heat and rain?

In the São Gabriel area the sun is overhead, or nearly overhead, at midday and it is very hot. Air pressure is low and the air is rising. Rainfall results from heavy convectional rainstorms in the afternoons. At this time evaporation and transpiration are greatest and it is very hot.

Figure D shows the rainforest water cycle. Throughout the Amazon region moist Atlantic trade winds are drawn in providing water for rainfall. The forest also has an important effect on rainfall. Up to 50% of all rainfall results from the recycling of moisture by evapo-transpiration. The issue of deforestation is a serious one. If the forest cover is lost then there will be less rainfall through the recycling of moisture (see pages 42–45).

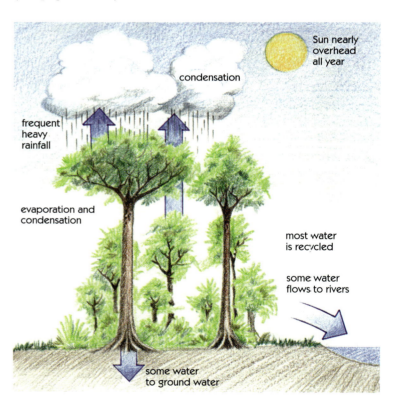

Sun nearly overhead all year

condensation

frequent heavy rainfall

evaporation and condensation

most water is recycled

some water flows to rivers

some water to ground water

Figure D: The rainforest water cycle

## Key ideas

● Where does the savanna climate occur?
● What is a savanna climate?
● Why has the climate a distinct wet and dry season?
● What influence does the climate have on the natural vegetation and people's lives?

## Main activity

Understanding and interpreting data

The savanna climate occurs between latitudes 5° and 15° both north and south of the Equator. It includes parts of West Africa including The Gambia and parts of South America including Brazil. The main characteristic of the savanna climate is its distinct wet and dry season. Temperatures are very warm throughout the year.

The climate graph, Figure A, is for Banjul in The Gambia. It is similar to other savanna locations such as northern Nigeria. Between June and October the rainfall is in heavy downpours often in the evening and night. It results from low pressure and convectional heating which cause warm air to rise and cool. Figure B shows why there is low pressure at this time between 5° and 15° north, which includes The Gambia. It also shows that between November and May there is high pressure over this area. The high pressure has sinking air which does not lead to rainfall.

Temperatures do vary although they are always warm and well above freezing. The warmest average temperatures occur in June when the sun is overhead but they dip during July and August as there is so much cloud. In December the sun is overhead at the Tropic of Capricorn (to the south), so the temperatures fall away. December and January are the coolest months and Gambians shiver and say the weather is cold. Minimum temperatures in Banjul can be as low as 15°C as the skies are clear of cloud and the heat of the day (over 30°C ) radiates to the atmosphere. At this time there is often a very dry and dusty wind from the Sahara regions known as the **Harmattan**.

The Gambian coast is cooler than places inland because of the cooling influence of the Atlantic Ocean. It is also wetter than places inland and to the north. There is evidence that rainfall totals throughout the West African savanna have been falling. The reasons for this may not be natural but the result of deforestation, over-cultivation, overgrazing and building. All these human activities take moisture out of the land and destroy natural vegetation, with the result that the atmosphere becomes drier, (Figure C). This process is known as desertification. Gambians are worried that this process will continue as their population rises from 1 million in the late 1990s to 2 million by 2025 and more and more demands are made on the land.

## The climate influences natural vegetation

Near the tropical rainforest where the rainfall is high and more evenly distributed, the savanna vegetation is woodland with some tall grass. Figure D shows that the vegetation

### Figure A: Climate graph for Banjul, The Gambia

Temperature °C

Rainfall (mm)

Average annual temperature 25.3°C
Average range of temperature 2.5°C
Annual precipitation 1320mm

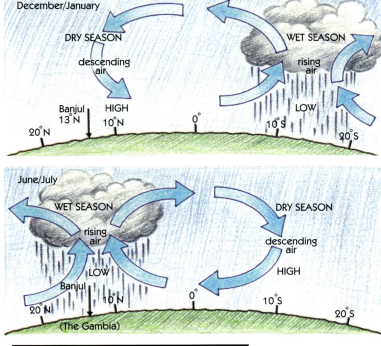

Figure B: Seasonal air circulation in The Gambia

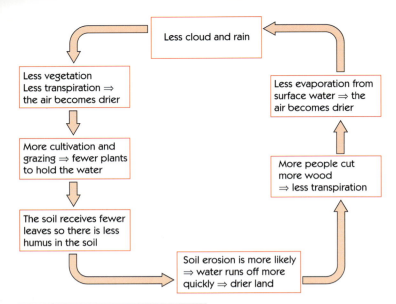

Figure C: The desertification process

Flowchart boxes:

Less cloud and rain

Less vegetation. Less transpiration ⇒ the air becomes drier

More cultivation and grazing ⇒ fewer plants to hold the water

The soil receives fewer leaves so there is less humus in the soil

Soil erosion is more likely ⇒ water runs off more quickly ⇒ drier land

More people cut more wood ⇒ less transpiration

Less evaporation from surface water ⇒ the air becomes drier

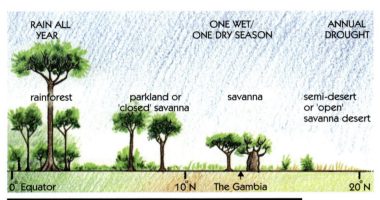

RAIN ALL YEAR — rainforest

ONE WET/ONE DRY SEASON — parkland or 'closed' savanna — savanna

ANNUAL DROUGHT — semi-desert or 'open' savanna desert

0° Equator  10°N  The Gambia  20°N

Figure D: The changing pattern of vegetation in The Gambia

slowly changes to more grassland with scattered trees. Figure E shows the savanna grass and forest at the end of the wet season. The landscape is green and luxuriant in the wet season and brown and parched in the dry season.

## The climate influences economic activity

The climate has traditionally dictated farming activities. Cereals such as upland or forest rice, millet and sorghum are planted at the beginning of the wet season and are harvested between October and December. More delicate vegetable crops such as tomatoes and peppers are planted at the end of the wet season to ripen in January and February. Figure F shows water being collected for young tomato plants in late October. The watering will have to continue throughout the dry season.

Since 1971 a tourist industry has developed in The Gambia. The dry season is favoured by tourists who come mainly from Western Europe. In the wet season many of the Gambians who work in tourism are without work. The climate is therefore an important influence on the tourist economy.

Figure E: Savanna grass and forest in the wet season

Figure F: Water is collected for young tomato plants in late October.

## ▼ Questions

1 Answer the following using Figure A and the text.
   a When does it rain in Banjul?
   b What is the total rainfall?
   c What and when is the highest temperature?
   d What and when is the lowest temperature?
   e What is meant by 'the average temperature'?
   f What minimum and maximum temperatures can be recorded in Banjul?
   g Why is it cooler on the coast of The Gambia?

2 a Why is there low pressure over The Gambia between June and October?
   b Why is there high pressure over the country between November and May?
   c What effect does this change of pressure have on rainfall?

3 Describe the natural vegetation that develops in the The Gambian savanna climate (Figures D and E).

4 How has the climate traditionally affected farming in The Gambia?

5 Write a report on the Gambian tourist industry for a new CD-ROM. It should be short, pointing out how the climate has encouraged the development of a tourist industry and how the climate has hindered the development of a year-round industry. ••

## Review

The Gambian savanna climate has a distinct wet and dry season because of the movement of the overhead sun which causes pressure systems to change. The climate influences farming practices and tourism.

# Human impact on climate

Many scientists think that the activities of people are having an increasing impact on the world's climates.

## Urban climates

The climate experienced by towns and cities is different from that of the surrounding countryside (Figure A).

- Average temperatures are 1°C higher
- Winter temperatures are up to 3°C higher
- Total precipitation is 5–10% greater
- Cloud cover is 5–10% greater
- Summer fogs are 30% more frequent
- Winter fogs are 100%.more frequent.

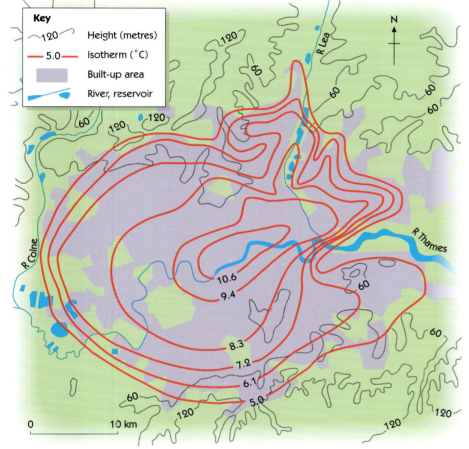

**Figure A:  The temperature pattern over London during a night in May**

There are three main reasons for these differences.

- The heating systems of the homes, offices and factories release heat into the atmosphere.
- During the day heat is absorbed by the bricks and concrete of the buildings and roads. This heat is released slowly during the evening and night.
- The blanket of urban pollution prevents the escape of some of the heat which would otherwise rise up into the atmosphere. There are ten times more dust particles found in the urban atmosphere. These particles provide the condensation nuclei needed for clouds to form and help to explain the increase in cloud, fog and precipitation over cities.

## Acid rain

Acid rain was first reported in the 1960s when Scandinavian scientists noticed that large numbers of fish were dying in the lakes and rivers of Norway and Sweden. They discovered that the acid content of the water had risen many times above natural levels. Further investigations revealed that the rainwater was acidic.

It did not take the scientists long to name likely culprits: Britain and Germany were singled out. The Scandinavians claimed that coal-burning power stations and oil refineries in those two countries produced the sulphur dioxide and nitrogen oxides which cause acid rain. Sulphur dioxide reacts with the atmosphere to form sulphuric acid; nitrous oxide develops into ozone and formaldehyde under the impact of sunlight. The prevailing south-westerly winds blew the acid rain over Scandinavia.

Soon afterwards the Germans themselves discovered that their lakes and forests were affected by acid rain.

Acid rain has several effects in addition to killing fish:

- it kills trees, especially conifers
- acid rain water percolates into underground water supplies
- crop yields fall in areas affected by it
- buildings crumble under its attack.

## Global warming

Average global temperatures have increased by 0.5°C during the twentieth century. The decade 1990–2000 was the warmest world-wide since reliable records began, over 150 years ago. Latest predictions are for an average increase of 3°C in global temperatures by 2030.

What is causing this? Some scientists still maintain that it is simply a normal increase caused by a natural cycle of warming and cooling. They point out that global temperatures were higher five thousand years ago than they are today. Most scientists believe that global warming is intensified by human actions. The burning of fossil fuels has released gases into the atmosphere which absorb heat rising from the Earth's surface and cause the atmosphere's temperature to increase, thereby intensifying the natural **greenhouse effect** (Figure B). Such gases are called greenhouse gases and include carbon dioxide, nitrous dioxide and methane. There are other causes of the release of greenhouse gases:

- Deforestation. Carbon dioxide is released through the burning of forests. Tropical rainforests are being burned at the rate of 11 million hectares per year.
- Vehicle exhaust emissions which release nitrous oxides.

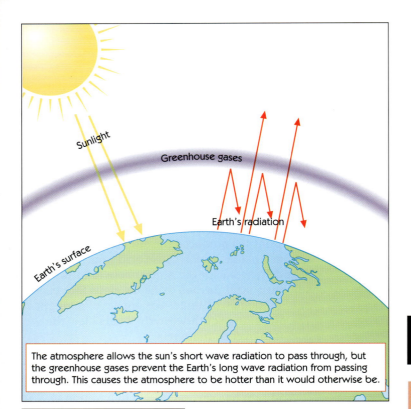

The atmosphere allows the sun's short wave radiation to pass through, but the greenhouse gases prevent the Earth's long wave radiation from passing through. This causes the atmosphere to be hotter than it would otherwise be.

**Figure B: The greenhouse effect**

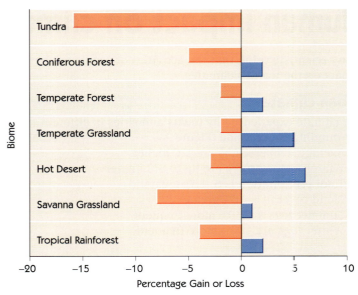

**Figure C: Estimated habitat change caused by global warming 2000–2080. All biomes except tundra will have some gain (blue bar) as well as loss (red bar)**

- The bacterial breakdown of organic matter. Humans have increased such emissions of methane by increasing herds of cattle, waste dumps and sewage treatment. Since 1980 world methane emissions have increased at an average of 10% per year.
  Global warming could have many dramatic effects.
- Widespread flooding may occur as polar ice caps melt.
- Increased rainfall, storms and winds could result as energy levels in the atmosphere increase.
- Disruption of world farming and food production would cause shortages and famine.
- The destruction of many fragile ecosystems and habitats: tundra, savanna grasslands and coniferous forests are especially vulnerable (Figure C).

It is difficult to be certain about the effects. The ocean currents would certainly be disrupted by the increased temperatures and this could result in some areas actually becoming cooler. If the Gulf Stream was affected the British Isles and Western Europe could become several degrees cooler on average.

Global warming is a challenge facing the world as a whole. Solutions will require international co-operation. In 1997 an international conference at Kyoto in Japan set targets for the reduction of greenhouse gas emissions. A second conference in Buenos Aires in 1998 began the next round of agreeing measures to cut emissions. This will not be easy because different governments have different priorities, particularly those of developing countries. Some poorer nations suspect that richer nations are deliberately using environmental arguments to prevent the poorer countries from developing their own industries.

## ▼ Questions

**1** Study Figure A.
  a This temperature pattern is called an **urban heat island**. Why do you think this term is used?
  b What was the temperature difference between central London and the surrounding countryside?
  c What causes this temperature difference?
  d Why does the concentric circular pattern break down towards the east?

**2** Imagine that you are a Scandinavian scientist working in the 1960s. You believe that fish are dying in Scandinavian lakes because of pollution from Britain. You have been given a two-minute slot on a British radio programme to explain your allegations. Decide on the key points you wish to make, write your script and then, if possible, tape record it.

**3** a Use a CD-ROM or textbook to help you to describe the Earth's natural greenhouse effect.
  b How are human actions affecting the greenhouse effect?
  c The table below shows the gases that intensify the greenhouse effect. Use a computer to draw a pie graph of these statistics.

| Gas | Percentage of Total |
|---|---|
| Carbon dioxide from fossil fuels | 40 |
| Methane | 18 |
| Carbon dioxide from deforestation | 15 |
| CFCs | 15 |
| Nitrous oxide | 5 |
| Others | 7 |

  d What are CFCs? What other important effect do CFCs have on the atmosphere? (Use a CD-ROM or textbook to help you answer this.)

**4** Design your own poster to highlight the possible effects of global warming.

# 4 Ecosystems

## Main activity

Research

## Do you know?

? The carbon cycle has been affected by the world's consumption of fossil fuels.

? The nitrogen cycle is a vital cycle in the biosphere.

An ecosystem is an area comprising plants and animals interdependent with their surrounding environment. The life cycles of the plants and animals are also closely linked to each other. An ecosystem may be small, such as a single pond, or it may be large, such as a tropical rainforest.

Two basic processes operate within ecosystems:
● flowing of energy
● cycling of materials.

## Energy flow

The sun is the main source of energy for all living things. Its energy is trapped by producers such as plants. They use the sun's energy, through the process of photosynthesis, to grow. The energy is passed on to animals which eat the plants (herbivores). The herbivores in turn may be eaten by meat-eating animals (carnivores) such as lions and foxes. This is a **food chain**.

Each link in the chain feeds on and obtains energy from the link below it, and in turn is consumed by the link above it. Each link is known as a trophic level (Figure A).

## Material cycling

Materials including chemicals such as carbon and nitrogen are circulated around the ecosystem in a continuous recycling process (Figure B).

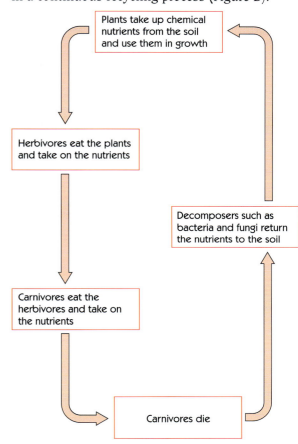

Plants take up chemical nutrients from the soil and use them in growth

Herbivores eat the plants and take on the nutrients

Carnivores eat the herbivores and take on the nutrients

Carnivores die

Decomposers such as bacteria and fungi return the nutrients to the soil

**Figure B: Material recycling**

| grass | worm | sparrow | sparrowhawk (or cat) |

| plankton | shrimp | cod | human |

**Figure A: Examples of food chains**

## ▼ Questions

1 What is an ecosystem?
2 a What is a food chain?
  b Give examples of food chains.
3 What is the relationship between **energy flow** and a food chain?
4 Use a CD-ROM encyclopaedia or a textbook to discover the following two material cycles affecting ecosystems:
  a the carbon cycle
  b the nitrogen cycle.

**Main activity**

Interpretation of a vegetation transect.

**Key ideas**

- Vegetation takes time to establish itself.
- A succession of vegetation can be seen in sand dune environments.

NONE (washed up seaweed only)

MARRAM GRASS

EAST

limited Gorse and Heather

beginning of sand dunes

sea (Studland Bay)

beach

sand dune - no real soil

A TRANSECT EAST – WEST FROM STUDLAND BEACH (approx. grid reference 038855)

scale: 2mm = 1m    0    5    10    15 metres

(Nb. predominant vegetation shown in capital letters)
LING = Calluna Vulgaris    BELL = Erica Cinera

© Crown copyright

**Figure D: Marram grass on new sand dunes**

**Figure E: Marram grass and heather**

**Figure A: OS map of Studland Bay**

**Key**
- Line of transect
- Mud flats at low tide

Line of transect

**Figure B: Sketch map of the sand dune area**

## Sand dune succession at Studland

At Studland Bay near Poole Harbour in Dorset on the south coast of England are several kilometres of **sand dunes**. Figure A is the extract from the 1:25,000 Ordnance Survey map showing the sand dune area. A sketch map showing the sand dune area is shown in Figure B, along with the line of a transect measured to show changes in vegetation. The location of the transect had to be chosen carefully because

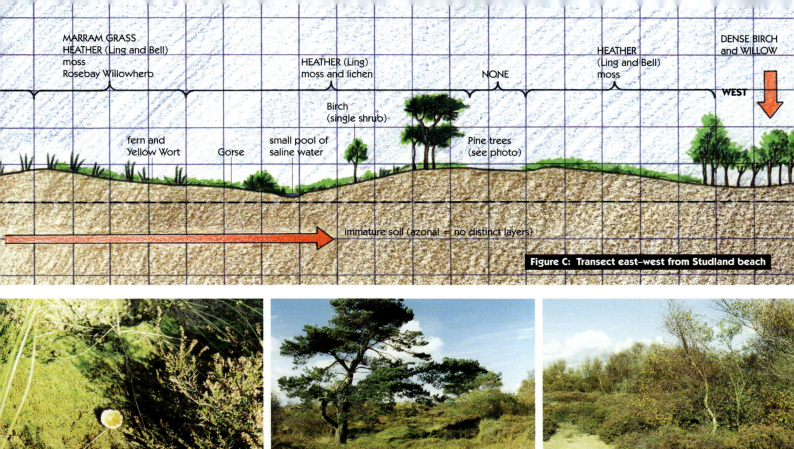

MARRAM GRASS
HEATHER (Ling and Bell)
moss
Rosebay Willowherb

HEATHER (Ling)
moss and lichen

NONE

HEATHER
(Ling and Bell)
moss

DENSE BIRCH
and WILLOW

WEST

fern and
Yellow Wort

Gorse

small pool of
saline water

Birch
(single shrub)

Pine trees
(see photo)

immature soil (azonal = no distinct layers)

**Figure C: Transect east–west from Studland beach**

**Figure F: Ground layer becomes more dense, with fungi**

**Figure G: Pine trees grow in denser vegetation**

**Figure H: Beginning of dense willow and birch woodland**

some areas were out of bounds – a National Trust project was protecting the endangered sand lizard. To measure a transect like Figure C, the fieldwork method is to use ranging poles, a tape measure and a clinometer which measures angles between the poles. The distances and angles are then drawn on graph paper using a protractor and ruler (the same transect method is used at Mappleton, see page 72).

The dunes have been formed by wind-blown sand which collects around obstructions such as tufts of grass or bits of wood. Some types of grass (see the transect, Figure C) can grow on the new sand dunes helping to stabilise them. Once the dunes have been established **marram grass** which is not so salt tolerant becomes the dominant grass (Figure D). It thrives on a continuous supply of fresh sand and produces underground runners from which new shoots grow. Once marram grass is established, the sand dunes can grow in height and there is less chance of the wind blowing them away. Even so rabbits, people, off-road motorbikes and very strong winds can still destroy what seems to be established sand dunes. This **ecosystem** is fragile.

On the more established First Ridge (numbered 2 on Figure C), the marram grass has decayed to give more organic matter called **humus**. The sand is developing into a soil and the vegetation is becoming a mix of marram grass and heath plants such as bell heather, ling and small gorse bushes. The soil becomes acid as the nutrients are washed down through the soil layers, the process of **leaching**.

Further inland the soil is more developed, and the vegetation begins to include trees and shrubs including birch, alder and Scots pine. See photographs E, F, G and H. This development of vegetation inland from the sea as the sand dunes become older is called a **vegetation succession**. In the case of Studland it is the result of changes in dune stability, salt content and soil formation over time.

### ▼ Questions

1 How are sand dunes formed?
2 What are the first plants to establish themselves?
3 Name the plants that can be found on the transect before trees grow.
4 What are the soils like along the transect?
5 Why does a vegetation succession exist in this area of sand dunes? ↦
6 How would you measure a transect?

### Review

As the Studland sand dunes have developed so a vegetation succession has become established. On the youngest dunes there is marram grass whereas on the older dunes there is dense birch and willow growth.

# Human impact on biomes

A biome is a large global ecosystem. Each biome takes its name from the dominant natural vegetation, such as tropical rainforest or savanna grassland (Figure A). The soils, plants and animals in each biome are broadly similar. Of course, human actions over the years have often greatly changed the biome. People may have cleared the natural vegetation, modified the soils and killed the wild animals.

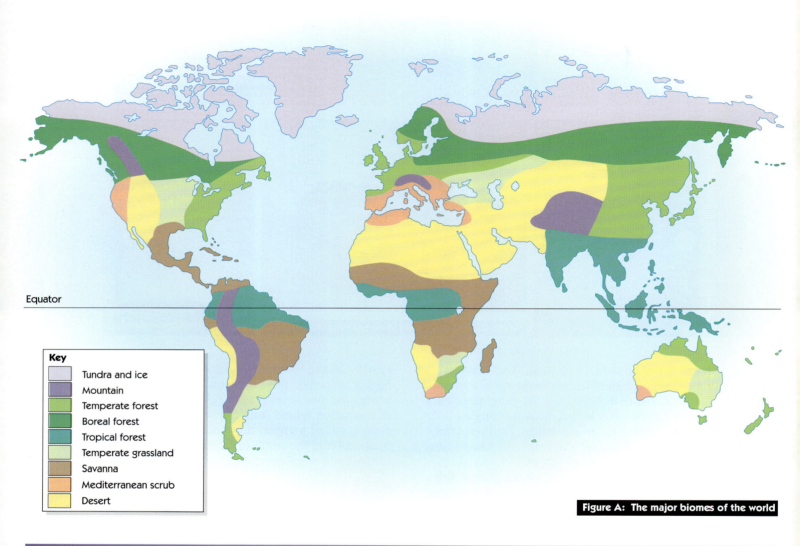

Equator

**Key**

- Tundra and ice
- Mountain
- Temperate forest
- Boreal forest
- Tropical forest
- Temperate grassland
- Savanna
- Mediterranean scrub
- Desert

**Figure A: The major biomes of the world**

## CASE STUDY: The coniferous forests of Sweden and Finland

Vast areas of forest stretch in an almost unbroken ring around the northern hemisphere of the Earth, from Canada through Scandinavia and northern Russia. This whole region has long, very cold winters and short, warm summers. Total annual precipitation is low (about 300mm). This region is known as the **taiga** or **boreal forest**.

The trees of the taiga are mostly evergreen and coniferous. They have developed characteristics which allow them to tolerate the climate and soils of the region:
- the needle-like leaves are thick and waxy to reduce water loss through transpiration; they are darkly coloured to absorb as much heat as possible
- cones shield the delicate seeds

- the bark is thick and contains resin which protects the trunk from the extreme cold and also from forest fires
- the conical shape of the tree prevents too much snow building up on the branches and breaking them off and also provides a degree of streamlining to protect against the strong winter winds
- the roots are shallow in order to absorb water from the less-frozen top soil.

In Scandinavia the coniferous forest has survived over vast areas because trees are often the most profitable crop that can be grown. There has been little pressure to create farmland since the climate is so harsh. Norway, Sweden and Finland have over one-third of Europe's forested land. In the most remote places the forest still survives in its natural state. Norway spruce and Scots pine are the dominant species. In other parts of Scandinavia the forests are managed; trees are felled for timber and pulp and paper. Sales of sawn timber have declined in importance as the demand has increased for chipboard and veneers. Demand for paper and newsprint, however, has increased dramatically. Forestry is vital to the Scandinavian countries; forest products account for over 10% of the value of exports of both Sweden and Finland, and 5% of Norway's.

Scandinavian forestry now provides a good example of sustainable forestry management. In the past trees were felled and not replaced. Since the 1970s much more care has been taken in the conservation of the forest. Larger areas are cleared, but sufficient trees are planted to replace those felled. Growth is speeded by transplanting young trees rather than growing from seed, and by using aircraft to spray fertiliser. Nevertheless, it still takes from 70 years in the south to 140 years or more in the far north for trees to reach maturity.

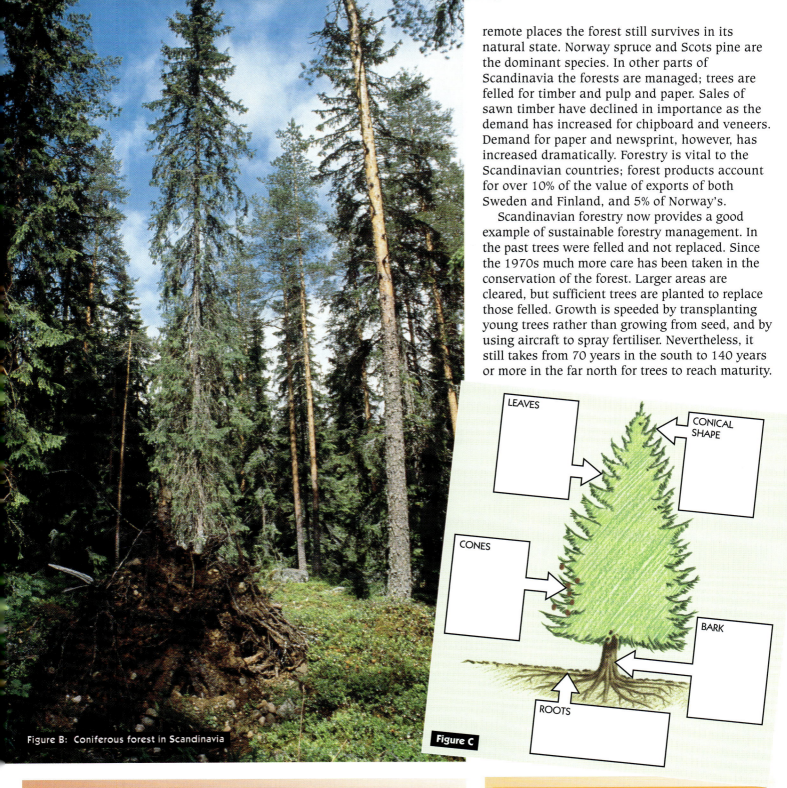

Figure B: Coniferous forest in Scandinavia

**Figure C**

LEAVES

CONICAL SHAPE

CONES

BARK

ROOTS

### Review

Large global ecosystems such as the tropical rainforest and the savanna are called biomes. The coniferous forest, or taiga, is the most extensive biome in northerly latitudes. The coniferous trees have developed characteristics which allow them to tolerate the harsh climate and poor, thin soils of the region. Human actions have greatly modified this biome and large areas are now exploited by the forestry industry.

### ▼ Questions

1. a  What is a biome?
   b  Name the nine major biomes.
2. What and where is the taiga?
3. Copy Figure C and add labels to show how the tree is adapted to the sub-Arctic climate.
4. a  What is meant by **sustainable** forestry?
   b  What do you think will be the advantages and disadvantages of switching to a sustainable form of forestry?

## Main activity

Decision making at a variety of scales.

## Key questions and ideas

● Where are the world's rainforests?
● What is a rainforest like?
● Why are rainforests important?
● There is widespread destruction of the rainforest.
● Rainforest people are threatened.
● Sustainable development of the rainforests should be possible.

The tropical rainforest is a major world biome. A biome can be defined as the interdependent community of plants and animals that evolve in a particular set of environmental conditions. It is estimated that the tropical rainforests produce 50% of the world's wood growth and yet cover only 8% of the world's surface. They are spread through three continents in a broad band around the Equator (Figure A).

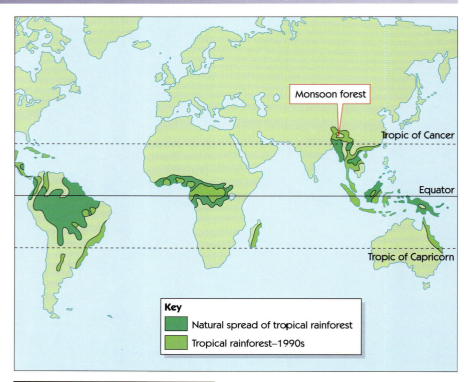

**Key**

| | |
|---|---|
| ▨ | Natural spread of tropical rainforest |
| ▨ | Tropical rainforest–1990s |

**Figure A: The world's tropical rainforest**

## A rainforest experience

A helicopter takes you from Manaus to the western Amazon near the River Negro (see page 32 for climate details). You spend a day with an experienced guide studying the rainforest ecosystem. The land around the river is often under floodwater and is called Varzea floodplain. The trees here have short trunks above flood level and many have aerial roots. It's a swampland and not the rainforest you expected. At least there are not many mosquitoes here as they do not like the acid water of the River Negro. Away from the floodplain the real rainforest or selva begins. It is stratified with at least four main layers. (Figures B, C and D illustrate the vegetation.)

Inside the forest it is difficult for you to realise the overall structure. It is very dark with only the odd ray of sunshine. The *ground layer* is a deep layer of old leaves, branches, twigs and the remains of dead insects and birds. There are termites, earthworms and fungi down here all living off this organic litter. It will all be re-cycled and the nutrients will be picked up by the shallow roots of the rainforest trees. Our guide dug a soil profile (see Figure E). Many people think that the rainforest soils must be fertile, and they are, while there is a layer of decomposing litter. But if that disappears the soil becomes very infertile. The nutrients get washed down the profiles (leaching). Sometimes the soil just gets washed away with the heavy rains (soil erosion). You were beginning to understand how the forest works – an input-output system like so much of physical geography.

The dark damp forest is alive with noises: there are chirps, cries, clicks, flaps, whines and hums. Most of the insects and birds are above you and only occasionally do you catch a glimpse of a colourful bird such as a toucan or hummingbird. Your guide shows you the footprints of an anteater and the skeleton of a sloth.

You can look up through the *understorey* to about 20 metres. This layer is made up of the trunks of the canopy trees as well as shrubs and plants. Most of the plants throughout the forest are evergreen. There is a large variety of plants unlike in other major biomes. This ecosystem is very dynamic and plants have adapted and specialised. Large fruit bats spread the seeds of fruit-bearing trees. Many small plants grow very slowly. When an emergent tree falls, sunlight finally gets through to the understorey. This is the one chance the plants need. They begin to grow upwards towards the light. Other plants have adapted their method of photosynthesis so they need very little sunlight.

Most of the rainforest is in the *canopy layer* 20–45 metres off the ground. If you could get into this layer you could see it was alive with birds and animals including reptiles. Most leaves are large so as to absorb as much sunlight as possible. One leaf can be the size of a person.

It is from the helicopter that the *emergent layer* can best be seen. Some trees reach up to 50–60 metres. These are the trees that have the big buttress roots seen at ground level. There are only a few of these giants in every hectare.

more light

shady
conditions

thin soil

buttress roots

shallow root
trees

Emergent Layer
trees grow very
tall - only a few
per hectare

Canopy Layer
trees with
climbing and
parasitic plants

Understorey
shrubs and
plants

Ground Layer
few plants, old
leaves, twigs,
decay, fungi

metres

60

40

20

0

Figure B: Layers of vegetation in the rainforest

Figure C: Fantastic rainforest plants

Figure D

## ▼ Questions

1 Describe the location of the tropical rainforests.
2 Read 'A rainforest experience'. Make a list of 20 words which help to describe
the structure of the forest.
3 With the aid of a simple diagram describe how the rainforest works.
4 Why are rainforest soils infertile once the forest has been cut down?

layer of
decomposing
leaves (litter)

decaying
leaves, fungi,
insects

rain dripping
to ground

poor soil -
few nutrients -
weathered rock

shallow root
systems

thin layer
of organic
rich soil

down to
parent rock

**Figure E: The soil under the rainforest**

**Figure F: Davi Kopenava Yanomami – spokesman for the Yanomami people, who travels the world to represent Amazonia's native peoples.**

## The importance of the rainforests

The Do you know? box gives some facts about importance of the rainforest for medicines. There are 80,000 species of tree and 55,000 species of flowering plants in Brazil, and the Amazon Basin contains 30% of all known plant and animal species. Rainforests store energy and moisture in their huge biomass (living matter). If this energy could be harnessed in a sustainable way then the whole world would benefit. The rainforests of the Amazon Basin are important for what they offer the world. Garo Batmanian, a conservationist from Brazil, said in 1997: 'We are not trying to put a fence around the Amazon or any other forest. OK, they are good for biodiversity, but they are good for people too, they make a good living out of them. All sorts of products can be harvested without destroying the forest.'

## Widespread destruction

It was estimated in 1997 that 65% of the natural rainforest area had been deforested. It is the rainforests of Asia that have been reduced most in percentage terms but the forests of South America have lost the largest area in terms of millions of hectares. Brazil has lost an area of rainforest the size of Spain in the last 20 years of the twentieth century.

Garo Batmanian points out: 'Five years after the forest is cut down the land is useless and the ranchers move on to new forest areas. We are proud of hosting the Earth Summit [Rio de Janeiro in 1994] and signed all the agreements but we did not keep our promises. The rate of forest destruction and the loss of species continues to accelerate.'

## The Yanomamo

The Yanomamo are a group of South American Indians that include many different tribes. They live in 200–250 villages in the Amazon Basin along the border between Brazil and Venezuela. They had little contact with the outside world until the 1940s. Their population is probably under 25,000. They are a hunting and fishing group of people and have suffered as the result of forest exploitation and destruction. There are fewer animals to hunt and fish have been killed by river pollution. In 1993 the *garimpeiros* (Brazilian gold prospectors) killed sixteen Yanomamo people.

Today the Yanomamo are helped by the Brazilian government and by non-government organisations (NGOs). The Brazilian government have set up the Yanomamo reserve north of the River Negro on the border with Venezuela. In 1994 a representative of the people (Figure F) spoke at the Rio Earth Summit to make the world aware of the plight of his people. In 1998 the world held its breath as fires raged in the north of Brazil. At one stage a Yanomamo leader radioed from the reservation saying that 'the fires are taking over our land, killing the animals we hunt and the birds in the trees.'

## Saving the forest

At all scales people are trying to manage the Amazon rainforests. Governments, non-government organisations (NGOs) and individuals are involved.

● 91 areas have been designated for protection. These vary in size up to thousands of hectares. Most are in the Brazil, Peru and Bolivia border areas.

## Sustainable development

The definition of sustainable development from the World Wide Fund for Nature (WWF) is 'improving the quality of life while living within the carrying capacity of supporting ecosystems'.

- The Brazilian government has promised to ban the extraction of mahogany trees from parts of the rainforest. This is difficult for a large country without resources to police the forests. Many large Asian multi-national companies now work in the South American rainforests.
- International Banks have started to negotiate with countries about debt repayments. They may drop demands for debt repayment if they promise to protect their rainforests.
- Private individuals have set up parks such as the Ecopark outside Manaus. Here the non-profit Living Rainforest Foundation promotes conservation and education. This is a form of **eco-tourism** where visitors pay $25 for a half-day excursion. More general tourism to fragile areas could be controlled.
- The buying and selling of hardwoods such as mahogany can be controlled by world organisations. Companies can agree only to buy mahogany from sustainable forests. Eastbourne council has replaced the town's groynes using sustainable hardwood from Guyana. Its decision was praised by many conservation bodies.
- There are ways of developing rainforest areas sustainably for wood and wood product production. **Agroforestry** imitates the canopy structure of the natural forest. Tree crops and short-lived farm crops are grown and local farmers plant tree seeds for future tree seedlings
- Local hunters and gatherers can be encouraged to collect tree products and are given a guaranteed market for them.
- Individuals can refuse to buy products from the rainforest but this is difficult unless they know what things contain forest products (see the Do you know? box below).

### Do you know?

? One-third of the world's remaining rainforests are in the Amazon Basin.
? 10 million living species are found in the rainforests.
? 10% of all non-prescription drugs are derived from rainforest plants.
? Curare, the poison used by indigenous Indians to tip their blowpipe darts, is widely used in operating theatres as a muscle relaxant.
? *Vinca rosea* has helped reduce the number of childhood leukaemia deaths.
? Rainforest Action Network (RAN) campaign to save the rainforest. They state:
? Between 1978 and 1996 more than 12.5% of Brazil's rainforests were destroyed.
? Industrial logging is the main cause of deforestation.
? Pulped rainforest trees go into toilet paper, cellulose products such as rayon, camera film and cigarette filters.
? 'Brazil is most at fault, but consumers and governments around the world have a role to play.'

### Review

The tropical rainforest is the world's richest biome but it is a delicately balanced ecosystem. There is a distinct structure to the rainforest. The rainforest is of global importance and its loss would be a disaster. Indigenous people such as the Yanomamo now receive protection. Saving the rainforest is a challenge that has to be tackled at a range of scales.

## ▼ Questions

5 Why does the Brazilian conservationist say, 'We are not trying to put a fence around the Amazon.'?
6 How much rainforest has been destroyed?
7 How have the indigenous Yanomamo indians suffered in recent times?
8 Write a full definition of sustainable development.

### Decision-making exercises

9 You are a conservationist in a meeting with the Brazilian Ministry of Environmental Affairs. What are your points of view and suggestions on the following three agenda items?
  a The problem of wandering gold prospectors in the Amazon basin who seem to have no regard for the forest or local indian people.
  b The problem of multi-national companies based in south-east Asia who are buying up large tracts of forest for future exploitation.
  c Private individuals who are setting up so-called eco-tourism centres in the rainforest.
10 Write an outline plan for conserving a large area of rainforest in the Amazon basin. You should assume the authorities agree to the proposals but have no resources to purchase nor police it.
11 You may work in small groups. Suggest how the following proposals would help reduce the destruction of the rainforests in the Amazon basin. Include reasons for your suggestions and examples from Brazil or elsewhere in the world.
  a A Brazilian government ban on the extraction of mahogany from the Amazon basin.
  b The setting up of core areas of forest where only research and the collection of genetic materials can take place.
  c A world conference to discuss the saving of the Brazilian rainforest.
  d The establishment of world agreements only to use tropical hardwoods that have passed a 'grown in a sustainable forest' test.
  e Extensive advertising in the developed world to show consumers that tropical hardwood is under threat and should not be purchased.

### Extension question

12 Research the importance of the rainforest and list the problems that will follow if it is all lost.

# 5

# The hydrological cycle

## Main activity

An exercise on a river in Snowdonia is presented as an enquiry to which students can respond, and on which they can base their own enquiry work. There are questions based on a 1:25,000 OS map.

## Do you know?

? Less than 1% of the world's water is fresh water available for human use.

## Key ideas

● The Earth's water circulates between the land, sea and air.
● Most of the water is stored in clouds, lakes, ice sheets and especially in the oceans.
● River discharge can be measured using simple equipment and simple calculations.

The hydrological cycle is the term used to describe our planet's water system – you may already have learned about it as the **water cycle**. Figure A shows that there is a constant movement of water between the sea, air and land.

● This movement is driven by heat from the sun which evaporates the surface waters of the oceans, lakes and rivers.
● The resulting water vapour rises into the air, cooling as it rises until it condenses to form clouds.
● Water may fall from the clouds in the form of rain, snow or hail (together known as **precipitation**).
● When water falls onto the land it may follow a number of routes back to the sea (Figure A):

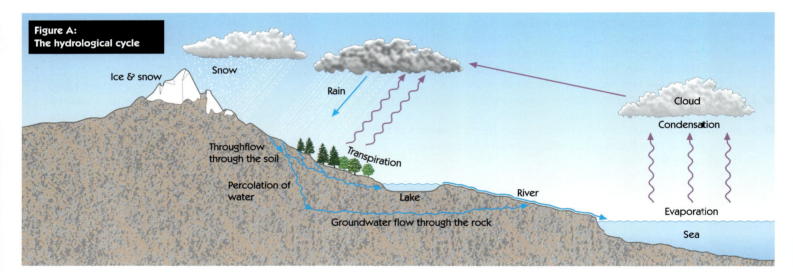

**Figure A:**
**The hydrological cycle**

Ice & snow

Snow

Rain

Cloud

Condensation

Throughflow through the soil

Transpiration

Percolation of water

Lake

River

Evaporation

Groundwater flow through the rock

Sea

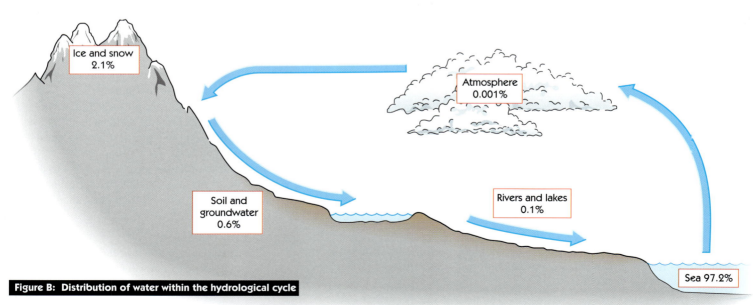

Ice and snow 2.1%

Atmosphere 0.001%

Soil and groundwater 0.6%

Rivers and lakes 0.1%

Sea 97.2%

**Figure B: Distribution of water within the hydrological cycle**

overland flow, throughflow, groundwater flow or may be stored in lakes or under the ground.

● Some of the water will be used by plants and released to the air by the process of transpiration.

The hydrological cycle is self-contained. There are no major inputs to nor outputs from the system. The amount of water within the cycle has been estimated as 1385 million cubic kilometres. This may sound a vast amount of water, but, as Figure B shows, almost all of the Earth's water is stored in the oceans. Only 2.8% is present in the atmosphere or on the land, and three-quarters of this is stored in the form of ice and snow. So only a tiny percentage is available for human use.

Rivers and lakes provide the major source of fresh water for human use. It is important for hydrologists to discover the amount of water available in rivers in order to plan for future water supply and also to be aware of any possible flood hazard. The following case study shows how you can calculate the **discharge** of a river using quite simple equipment.

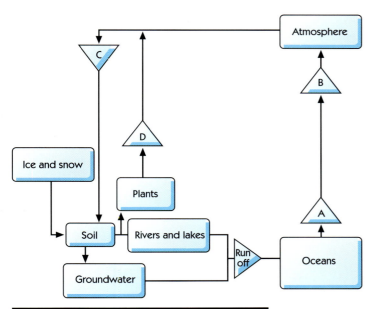

**Figure C: Systems diagram of the hydrological cycle**

## CASE STUDY: The Afon Nant Peris, Snowdonia

The mountains of Snowdon and Glyder Fawr tower over the narrow Pass of Llanberis in Snowdonia, North Wales. A small river, the Afon Nant Peris, flows along the pass and forms the subject of this case study which takes the form of an enquiry.

### The Afon Nant Peris Enquiry

**Aim**

to investigate:

● the dimensions of the river channel
● the velocity of the river
● the calculation of the river's discharge
● the following question:
  Do velocity, discharge, channel width and depth increase with increasing distance downstream from the river's source?

**Methods**

● Visit the river and conduct fieldwork measurements at a selection of sites along the river
● Observe the sites, take photographs and make sketches
● Calculate the discharge using a simple formula

**Results**

● Present the data in a variety of forms
● Attempt to identify any relationships between velocity, discharge, channel width and channel depth and distance downstream from the river's source.

## ▼ Questions

1 Make a copy of Figure C and fill in the missing labels A, B, C and D.

2 Use Figure A to find the words which mean:
  a water vapour turning to water droplets
  b water seeping down into the soils and rocks
  c water moving through the soil
  d water vapour given off by plants.

3 What percentage of the world's water is found
  a in the oceans
  b in the atmosphere?

4 a What is the meaning of the term *discharge* of a river?
  b Give two reasons why hydrologists need to know the discharge of rivers.

Figure D is an OS 1:25,000 map which shows the source of the Afon Nant Peris close to the youth hostel at grid reference 647557. The channel of the river first appears in the marshy land beside the A4086, fed from many rills running down the steep mountainsides on either side of the pass. An average of 2500mm of rain falls each year on these mountains and 80% of it flows along the river. Figure E shows the river near its source, at this point a tiny stream flowing down the impressive glaciated valley.

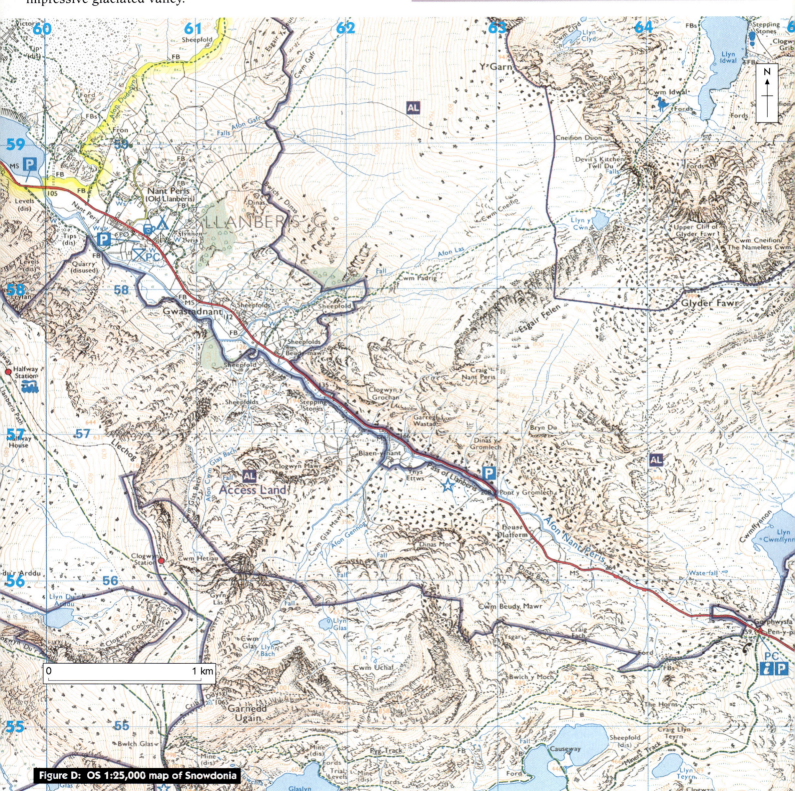

**Figure D: OS 1:25,000 map of Snowdonia**

© Crown copyright

**Figure E: Near the source of the Afon Nant Peris**

Figure F shows the course of the Afon Nant Peris from its source to the point where it flows into a lake, Llyn Peris. This is the longitudinal, or long, profile. As with most rivers, Afon Nant Peris has an uneven long profile. The valley was glaciated during the Ice Age and two lakes occupy part of the valley. Our fieldwork measurements were conducted on the river above the lakes at the four sites located on Figure F. Figures G–J show the four sites.

▼ **Questions**

**1** Study the OS 1:25,000 map extract (Figure D).
a Name the tributaries which join the Afon Nant Peris at grid reference: (i) 614575 (ii) 613577.
b What is the height at 629565?
c What boundary does the Afon Nant Peris cross at 609587?

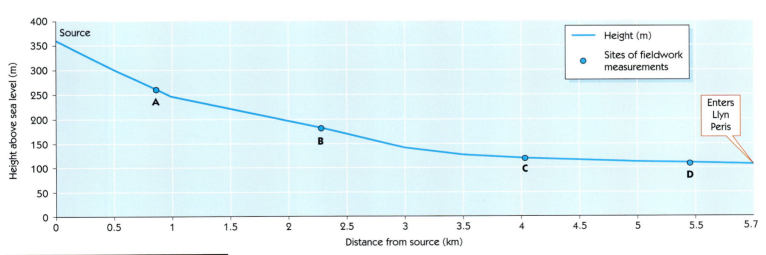

**Figure F: Long profile of the Afon Nant Peris**

## The enquiry assignment

1 To calculate the discharge of the Afon Nant Peris from the fieldwork measurements provided, with the assistance of the worked example (Figure K).
2 To construct graphs to show the relationship between distance downstream and discharge, velocity, channel width and depth.
3 To conduct fieldwork measurements of a stream in order to calculate its discharge.

### Method

#### Fieldwork measurements

1 Choose a straight section of the stream with an even width and measure a 10m distance along the bank.
2 Measure the width of the stream at the water level.
3 Using a metre ruler, measure the depth of water at 50cm distances across the stream.
4 Drop an orange into the stream above the first marker pole. Note the time, in seconds, that the orange takes to travel the 10m. This should be repeated ten times in order to obtain an average time. The orange should be dropped into the water at different points across the stream.

#### Discharge calculations

1 Calculate the cross-sectional area. This is calculated using the following equation: cross-sectional area = channel width × average channel depth
2 Add up the ten times taken by the orange to travel the 10m distance and calculate the average time.
3 Divide 10 by the average time: this gives you the average velocity in metres per second.
4 Multiply the velocity by the cross-sectional area: this gives you the discharge in cubic metres per second (cumecs).
5 Multiply the discharge by 0.8 (the orange floats just beneath the surface where the stream's velocity is higher than the average velocity for the whole depth of the stream. Multiplying the discharge by 0.8 gives you a more accurate result for the average discharge of the whole stream).

Figure H: Site B

Figure I: Site C

Figure G: Site A

Figure J: Site D

## Analysis

1 Study the worked example (Figure K) carefully. Use it to help you calculate the statistics needed to complete a copy of the table below.

SUMMARY TABLE

| Site | Width (m) | Average depth (m) | Cross-sectional Area (m2) | Average time for orange to travel 10 m | Velocity (m/s) | Discharge (cumecs) | Discharge (cumecs) x 0.8 |
|------|-----------|-------------------|----------------------------|-----------------------------------------|----------------|--------------------|---------------------------|
| A | 2.7 | 0.33 | 0.89 | 57.5 | 0.174 | 0.155 | 0.155 |
| B | 3.1 | | | | | | |
| C | 6.8 | | | | | | |
| D | 7.1 | | | | | | |

2 a Why were ten timed runs of the orange needed?
  b Why was it important to drop the orange at different points across the stream?
  c Why was it necessary to multiply the discharge figure by 0.8?

### WORKED EXAMPLE

**The following calculations are based on measurements taken on the Afon Nant Peris in May 1998 at Point A (Figure L).**

**Discharge calculations**
- Average channel depth = 2.28 m / 7 = 0.33 m
- Cross-sectional area = 2.7 x 0.33 = 0.89 square metres
- Average time for orange to travel 10 metres = 575 / 10 = 57.5 sec
- Average velocity = 10 / 57.5 = 0.174 metres per second
- Discharge equals Velocity × Area = 0.174 × 0.89 = 0.155 cumecs
- 0.8 × Discharge = 0.155 × 0.8 = 0.124 cumecs

**Figure K: Worked example**

**Fieldwork measurements**

| | |
|---|---|
| Width of channel: | 2.7 |
| Gradient of channel: | 5.0 |

Depth of channel:

| Distance from west bank (cm) | 0 | 0.5 | 1 | 1.5 | 2 | 2.5 | 2.7 |
|---|---|---|---|---|---|---|---|
| Depth of water (cm) | 0.21 | 0.25 | 0.35 | 0.44 | 0.19 | 0.39 | 0.45 |

Recorded times for the orange over the 10 metre distance:

| Run | 1 | 2 | 3 | 4 | 5 | 6 | 7 | 8 | 9 | 10 |
|---|---|---|---|---|---|---|---|---|---|---|
| Time (seconds) | 65 | 53 | 56 | 54 | 56 | 63 | 59 | 70 | 54 | 45 |

**Figure L: Fieldwork measurements on the Afon Nant Peris**

## Conclusions

For both of the following questions, ensure that you use all the resources provided including the map (Figure D) and the five photographs (Figures E, G, H, I and J).

1 Figure M is a graph showing how channel width varies with distance downstream.
  a What relationship does the graph show?
  b Draw your own graphs for distance downstream v average depth, velocity and discharge. What relationships do your graphs reveal?

2 The measurements were recorded during May. How do you think the results might differ if they were recorded during
  a September
  b February?

3 Now choose your own section of a river and calculate the discharge. Try calculating the discharge several times over the period of a year; you will gain a useful insight into the changing nature of the river's discharge.

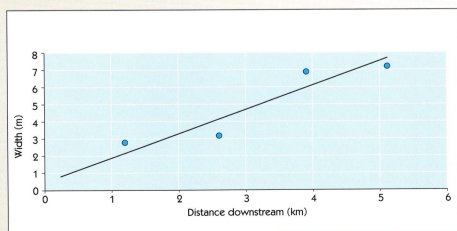

**Figure M: Graph of distance downstream v channel width**

# The water balance

## Main activity

Students develop an understanding of the factors affecting storm hydrographs, and draw hydrographs to show the effects of different parameters.

## Key ideas

● The water balance is the relationship between the input and output of water in a drainage basin.
● A **storm hydrograph** shows the response of a river's discharge to a storm.
● Storm hydrographs can be affected by many different factors which change their shape.

## The water balance

The **water balance** of a river is the relationship between the amount of water entering the drainage basin through precipitation and the amount leaving it in the river. The difference between the two is called the loss. Water loss occurs through evaporation, transpiration by plants and storage. We have seen how to measure the discharge of a river. **Evapotranspiration** can also be measured and the water balance calculated as explained in Figure A.

The water balance in the UK is usually calculated over a period of 12 months running from 1 October to 30 September. This is called the **water year**. This period is chosen because the end of the summer is the time when least water is stored within the drainage basin. The water balance for the Afon Nant Peris has been calculated over a year as shown in the box below.

> Precipitation 2480mm = Discharge (1970mm)
> 1 – evapo-transpiration (510mm) + storage (0)

A very high percentage (almost 80%) of the precipitation entering the drainage basin of the Afon Nant Peris leaves as discharge. In eastern England the percentage may be lower than 20%.

**1 RAINFALL** is measured using a **rain gauge**. The water collected in the bottle is poured into a measuring cylinder which registers rainfall in millimetres. In order to gain an accurate figure for rainfall several rain gauges must be used across the drainage basin. If enough gauges are used, isohyets (lines joining points of equal rainfall) can be drawn on a map of the drainage basin.

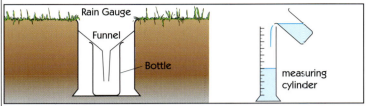

**2 EVAPOTRANSPIRATION** can be measured using a lysimeter. A column of soil is cut and placed inside a container such as an oil drum. The container is then inserted into the hole in the ground. The rainfall at the site is measured and the flow of water through the lysimeter is trapped in the lower part of the can (see diagram). The amount of water stored within the soil can be calculated by weighing the upper can (1ml of water weighs 1g). The amount of evapotranspiration can then be found by using the following formula:

Output (the amount of water collected in the lower can) = rainfall – evapotranspiration – storage

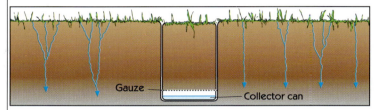

**3 RIVER FLOW (DISCHARGE)** can be measured using a **current meter**. This measures the velocity of the river. The amount of water passing through the river channel (the **discharge**) can be calculated using the following formula:

$$discharge = velocity \times cross\text{-}sectional\ area$$

It is rather more fun to measure discharge by using an orange (see page 50). More sophisticated methods of measuring discharge include weirs and flumes. Flumes are artificial channels of known cross-sectional area built into the river bed. Since the area is known only the depth of water needs to be obtained in order to calculate the discharge. Some flumes are continuously monitored by computer.

**4 THE WATER BALANCE** can now be calculated using the following equation:

$$Discharge = precipitation - evapotranspiration \pm storage$$

**Figure A: Measuring the water balance of a drainage basin**

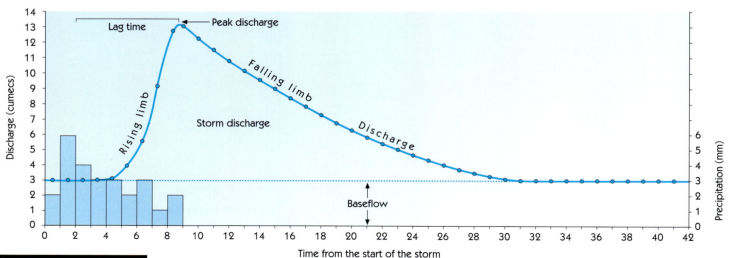

**Figure B: A storm hydrograph**

# The storm hydrograph

Figure B is an example of a storm hydrograph. The discharge of a river is plotted against time from the start of a storm. You will notice that the river's discharge increases in response to the input of water from the storm. There will be a significant delay between the peak of the rainfall and the peak discharge (the lag time) because it will take some time for the water to reach the river.

The overall shape of the storm hydrograph will vary from river to river because there are many individual factors involved, such as:

● the intensity of the storm
● how long the storm lasted
● the shape of the river basin
● whether the river basin was dry or saturated before the storm began

## ▼ Questions

**1** a What percentage of the rainfall in (i) Snowdonia (ii) eastern England runs off in the rivers?
  b Why do you think these two values differ so much? (Think about the effects of different climates, rock types, slopes and land use).
**2** Sketch two hydrographs showing how a river's discharge would differ if the soils of the river basin before a storm began were
  a dry
  b saturated.

● the rock type of the river basin
● the land use of the basin, and type of vegetation.

Figure C shows how the river hydrograph might be affected by some of the factors listed above.

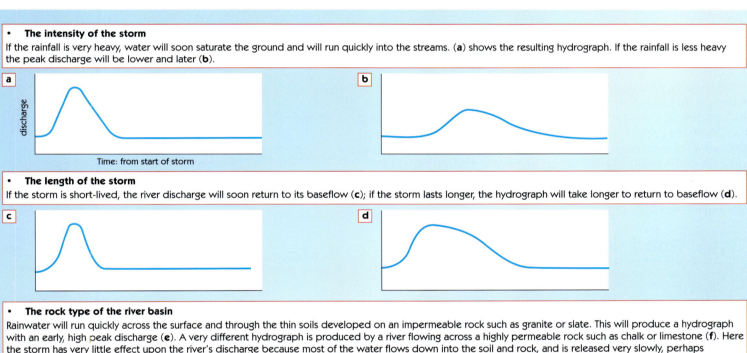

> **• The intensity of the storm**
> If the rainfall is very heavy, water will soon saturate the ground and will run quickly into the streams. (**a**) shows the resulting hydrograph. If the rainfall is less heavy the peak discharge will be lower and later (**b**).

a (discharge / Time: from start of storm)    b

> **• The length of the storm**
> If the storm is short-lived, the river discharge will soon return to its baseflow (**c**); if the storm lasts longer, the hydrograph will take longer to return to baseflow (**d**).

c    d

> **• The rock type of the river basin**
> Rainwater will run quickly across the surface and through the thin soils developed on an impermeable rock such as granite or slate. This will produce a hydrograph with an early, high peak discharge (**e**). A very different hydrograph is produced by a river flowing across a highly permeable rock such as chalk or limestone (**f**). Here the storm has very little effect upon the river's discharge because most of the water flows down into the soil and rock, and is released very slowly, perhaps sustaining the river for weeks or even months after the storm occurred.

e    f

> **• The land use of the basin: type of vegetation**
> (**g**) and (**h**) show how different types of vegetation can produce different hydrographs. The early, higher peak of (**g**) is due to the pastureland having little effect in slowing down the movement of the water. The later, lower peak of (**h**) is due to the trees intercepting much of the water and slowing its progress downslope.

g    h

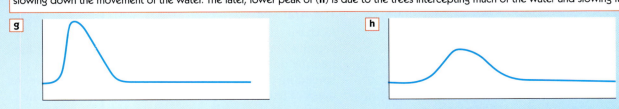

**Figure C: Factors affecting the shape of the storm hydrograph**

# Human impact on the hydrological cycle

## Main activity

Students can develop an understanding of the varied effects of human activities on the hydrological cycle. Main activities include textual and data analysis, and the drawing and interpreting of graphs.

## Key ideas

● Human activities can greatly affect the natural hydrological cycle, at a variety of scales.
● Human activities can cause flooding and create drought.

People can have major impacts on the hydrological cycle at all scales.

## The global scale

The burning of fossil fuels is thought by many scientists to be responsible for global warming which is creating increased evaporation and increased rainfall (see pages 35–6).

When forests are removed, transpiration is reduced and that reduces total rainfall. Deforestation has also resulted in more rapid throughflow and overland flow causing flooding, such as the example from Thailand described in Figure A. In November 1998 over 24,000 lives were lost in central America when Hurricane Mitch delivered the equivalent of a year's rainfall in just one week. The torrential rain caused mudslides which engulfed villages and swept shanty homes from river banks and floodplains. The intensity of the flooding was again made worse by deforestation.

### Thai Flood Caused by Over-logging

A vast heap of mud, sand and timber is all that remains of Katoon, a farming village of 300 families, after devastating floods and landslides in southern Thailand last week.

Katoon was obliterated in half an hour when days of heavy rain brought flash floods and dislodged mud and sand from hills overlooking the valley. The mudslides sent boulders and thousands of felled timber logs crashing into flimsy wooden houses, destroying the already flooded village and killing at least 168 of the villagers. Villagers blamed the tragedy on illegal logging. This eliminates natural shade needed for the growth of ground foliage that traps rainfall in the highlands during the monsoon. Deep red gashes mark the mountainsides of the once fertile valley of rice fields and orchards, now a wasteland metres deep in mud and logs.

**Figure A: The effects of deforestation in Thailand**

## The regional scale

### Southern California

The construction of reservoirs and irrigation schemes can alter the hydrology of drainage basins. Southern California's vast urban areas are supplied with water by aqueducts which carry water hundreds of kilometres from dams in the Rocky Mountains (Figure B). Southern California is a semi-desert area (San Diego has an average annual total of 240mm rainfall). Its fruit and vegetable farms, plus its growing urban population, are almost entirely dependent upon the water supplied by this huge water transfer project.

Supplying water to southern California has created problems:
● A large percentage of the water is lost through evaporation. This increases the salt content (salinity) of the water and adds salt to the soil, reducing crop yields.

**Figure B: The Southern California water transfer**

● The salt can only be removed from the soil by flushing it away with large amounts of fresh water.
● The reservoirs slowly fill with sediment brought by rivers flowing into them.
● The high construction costs mean that farmers' water bills are high, so increasing the prices they have to charge for their produce.

### The Aral Sea

In Uzbekistan a vast irrigation project has enabled cotton to be grown in a desert. However, the project has had the unintended effect of shrinking the Aral Sea, spreading disease and changing the regional climate. A mass of irrigation canals were built to drain away the water from the two major rivers feeding the Aral Sea. So much water is taken from these rivers that hardly any now enters the lake.

The Aral Sea was the fourth largest lake in the world, but since 1960 it has lost over 40% of its surface area. This inland sea has no outlet except evaporation. The irrigation projects have destroyed the balance between inflow and evaporation. As the amount of fresh water supplying the lake has been cut, the evaporation has continued unchecked, causing the Aral Sea to shrink and become more salty. Experts estimate that the sea will disappear completely within the next thirty years.

Salt flats now mark the former lake bed. Winds whip up the salt and carry it away in great dust storms to harm the land – and the people – of the region. The salt builds up in the soil and poisons the crops. People breathe in the salt and drink the polluted water supplies. There has been a great increase in eye and respiratory diseases, and throat cancers.

Figure C: The shrinking Aral Sea

The shrinking of the Aral Sea has affected much of central Asia. Salt has been blown as far away as the coast of the Arctic Ocean. The lake has until recently had a moderating influence on the climate of the region, but now the climate has become more extreme. One result is that the growing season for crops is now shorter.

## The local scale

Changing land uses can have major effects. For example, the construction of urban areas creates impermeable surfaces such as roads, car parks and pavements. Rainwater flows quickly into the drains and sewers and is fed rapidly into the streams and rivers, causing increases in discharges and possibly flooding. The construction of a new town in Essex affected the storm hydrograph of a river in the area. Harlow was designated a new town in 1947 at which time the village of Harlow had a population of 4500. Progress was slow at first, but rapid urban development soon got under way and by 1961 the new town had a population of 53,000. By 1971 it had reached 77,000.

Figure D: Fishing boats abandoned on the Aral Sea

## Review

Human activities can greatly affect the natural hydrological cycle, at a variety of scales. Irrigation schemes can allow the desert to bloom, but they can also bring problems. In southern California, increased salinity of the soils reduces yields. Much more serious effects have resulted from the irrigation projects in Uzbekistan – the Aral Sea has shrunk by almost half of its surface area since 1960. In other locations human actions have caused flooding or drought.

## ▼ Questions

1  Read the article (Figure A).
  a  Where was the village of Katoon?
  b  What happened to the village?
  c  How did deforestation contribute to this disaster?
  d  How does the climate of the region help to explain the extent of the flooding?
  e  'Deep red gashes mark the mountainsides of the once fertile valley.' What do you think the deep red gashes are and how were they caused?

2  a  What is meant by a **water transfer project** such as that supplying Southern California?
  b  Why is this project so important to Southern California?
  c  What problems does the project cause?

3  a  Where is the Aral Sea?
  b  Using Figures C and D to help you, explain what has happened to the Aral Sea.
  c  Draw two line graphs using the data in the table below:

| Year | 1930 | 1940 | 1950 | 1960 | 1970 | 1980 | 1990 |
|---|---|---|---|---|---|---|---|
| Irrigated area (million hectares) | 3.9 | 4.2 | 4.8 | 5.1 | 5.7 | 6.0 | 7.0 |
| Surface height of the Aral Sea (metres above sea level) | 53 | 53 | 53 | 53 | 51 | 46 | 41 |

  d  Why has the flow of water into the Aral Sea been greatly reduced?
  e  How and why has the shrinking of the Aral Sea changed the climate of the region?

# 6 River processes

## Main activity

Understanding the processes which contribute to the discharge of a river.

## Key ideas

● The processes which supply water to a river operate at different speeds, contributing different parts of the discharge shown by the storm hydrograph.
● Rivers erode, transport and deposit material through a number of processes.

## The processes leading to stream flow

Rivers begin to flow when water cannot infiltrate into the ground, but flows across it. This overland flow can create tiny channels called rills. The rills grow into brooks, streams and rivers.

● **Overland flow** occurs when rain hits the ground's surface faster than the ground can absorb it. River flow is also fed by water coming from other sources.

● **Channel precipitation**: rainfall directly into the river. This is only a tiny proportion of the total rainfall, but water from this source is immediately available for river flow and is not subject to immediate loss.
● **Throughflow**: rainwater infiltrates the soil and then flows downhill within the soil.
● **Groundwater flow**: water penetrates beyond the soil into the rock below and makes its way down to the water table (the top of the saturated zone within the rock). The water can then flow very slowly through the rock and emerge at the surface as a spring.

## River processes

Rivers erode, transport and deposit material.

### Erosion

Figure B shows the Afon Nant Peris in Snowdonia. The water rushing over the rocky bed is eroding material in three ways.

● The running water itself carries away the loose material. This process is called **hydraulic action**.

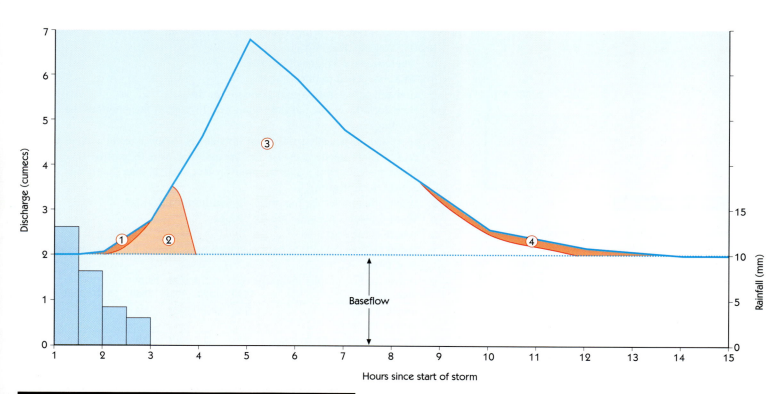

**Figure A: Flood hydrograph divided into the contributory processes**

**Figure B: The Afon Nant Peris, showing the size of boulders which can be moved only in times of flood**

- The rocks and pebbles carried by the river crash against the sides and bed of the channel and remove more material. This process is called **corrasion**.
- The rocks and pebbles carried by the river crash into each other and break up into smaller fragments. This is called **attrition**.

River water can also dissolve the minerals in some rocks, for example the calcium carbonate in chalk and limestone. This is the process of **solution**.

The appearance of the river valley is largely due to the action of the river itself, and to the weathering on the slopes above the river. The amount of erosion depends upon the speed of water flow, the amount of material carried and the type of rock over which the river flows. Soft rocks are eroded much more quickly than strong and resistant rocks.

Most of the time there is little erosion taking place. Most erosion happens in times of flood. The increased input of water gives the river much more energy. The velocity and discharge will greatly increase and the river will be able to carry much more material. Even large boulders can be swept along, crashing against the bed and banks of the channel (Figure B). As the river loses energy, the flood subsides and a good deal of material is deposited by the river.

## Transportation

Rivers transport material in four main ways:
- the largest material is pushed or rolled along the river bed. This process is called **traction**
- smaller stones and pebbles are bounced along the river bed in a process called **saltation**
- some particles such as silt and clay are small enough to be carried along within the water in **suspension**
- some minerals such as calcium carbonate are dissolved and carried along within the water in **solution**.

## Deposition

Once the river's velocity decreases it will begin to deposit material. The larger material will be deposited first and the finer material will be carried much further since it needs a lower velocity to transport it than it did to pick it up. This results in a grading, or sorting, of river deposits by size.

### ▼ Questions

1 When does overland flow occur?
2 What is meant by
  a channel precipitation
  b throughflow?
3 Study the flood hydrograph (Figure A). Which processes do you think are indicated by the areas numbered 1–4? (Think about the relative speeds of the four processes.)
4 Write a sentence to explain each of these processes of erosion:
  a attrition
  b corrasion
  c solution.
5 List three factors affecting the amount of erosion in a river valley.
6 Write a sentence to explain each of these processes of deposition:
  a traction
  b saltation
  c suspension
  d solution.
7 Why is material deposited by a river sorted by size?

## River landforms

The Afon Nant Peris is a short river, only a little over 6km in length, yet in that short distance it has developed examples of several of the main landforms resulting from river processes.

In its upper section the river is narrow and its bed is strewn with large stones and boulders (Figure A). In places there are steeper sections called rapids (Figure B); the faster flowing 'white water' marks points where the rock is more resistant and the river's velocity increases.

Further down the valley the river is broader and deeper (Figure C). The large boulders and stones have been worn down by erosion. Notice how rounded many of the stones are beside the river.

Figure D shows a bend, or meander, on the lower section of the Afon Nant Peris. The meander has a clearly defined river cliff on the outside of the bend and a river beach, also called a slip-off slope, on the inside. Figure E shows how the main force of the river's flow is concentrated on the outside of the meander. Erosion is therefore greatest on the outside of the meander; the bank is undercut to form a steep river cliff. On the inside of the meander the water is shallower and slower moving, so some deposition occurs there, forming the river beach.

Figure A: Boulder strewn bed of the Afon Nant Peris in its upper section

Figure B: Stretch with small rapids

Figure C: Lower section of the Afon Nant Peris

Figure D: Part of a meander on the Afon Nant Peris

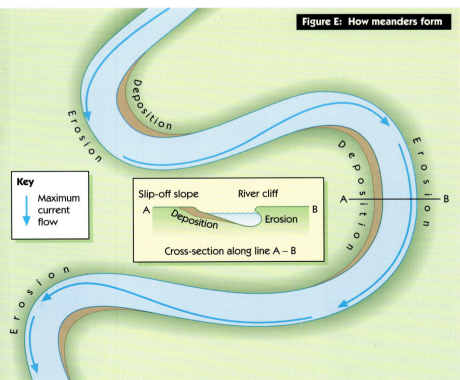

Figure E: How meanders form

Key

Maximum current flow

Deposition

Erosion

Slip-off slope    River cliff

A    Deposition    Erosion    B

Cross-section along line A – B

A —————— B

Erosion

Deposition

Erosion

Deposition

Deposition

Erosion

Erosion

## ▼ Questions

1  a  Compare the two stretches of river shown in Figures A and C.
   b  Explain how and why the size of rocks at the two sites differs.
2  Using Figures D and E, explain how meanders are formed.
3  Using a Geography textbook or CD-ROM discover how some meanders can become ox-bow lakes. Draw an annotated sketch diagram to illustrate the formation of an ox-bow lake.

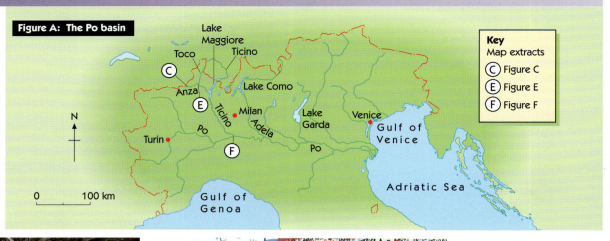

Figure A: The Po basin

Lake Maggiore
Lake Ticino
Toco
Ⓒ
Anza
Ⓔ
Ticino
Lake Como
Milan
Adela
Lake Garda
Venice
Gulf of Venice
Turin
Po
Ⓕ
Po
Adriatic Sea
Gulf of Genoa

N

0    100 km

Key
Map extracts
Ⓒ Figure C
Ⓔ Figure E
Ⓕ Figure F

Figure B: The mountain stream at point A on the map

Figure C: Part of Italian 1:50,000 map

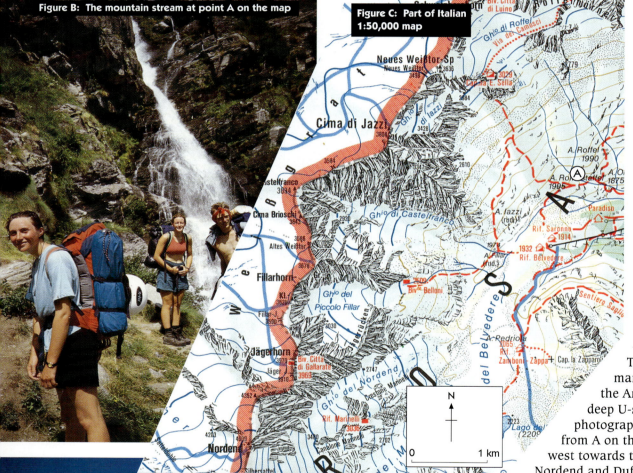

The river Po is in north Italy; its basin is shown in Figure A. In the Po basin are several river landforms, ranging from the glaciated valleys of the Alps to the meandering lowlands of the plain and the delta at the Adriatic Sea.

The source of some of the Po's waters are high in the Italian Alps near Monte Rosa on the Italian–Swiss border. Streams of water fall down the steep valley sides as shown in Figure B. This photograph was taken at A on the Italian 1:50,000 map Figure C. The map shows that many of the **headwaters** of the Anza river are glaciers in deep U-shaped valleys. The photograph, Figure D, was taken from A on the map looking south-west towards the mountains of Nordend and Dufour (beyond).

### ▼ Question

1 Study the map and the photographs of the Anza river in the Italian Alps.

a Describe the landscape in terms of its rivers, streams, glaciers and valleys.

b How are the contours shown on the glaciers?

c What evidence can you find to show this is (i) a skiing area (ii) a walking area?

Figure D: On the upper section of the Anza valley

The Anza river flows east in a steep-sided glacial valley, which has numerous tributaries, until it reaches the river Toce. The confluence of the two rivers is shown on the 1:50,000 map extract, Figure E. The Anza meets the Toce with many distributaries making it look like a delta. South of Vogogna the river channel is **braided** (it splits up with small islands of deposit). The valley of the Toce has developed a floodplain which is wide compared to that of the Anza.

The river Toce flows into Lake Maggiore which is a large glacial ribbon lake which forms part of the river Ticino. The river Ticino is larger than the Toce with a wider floodplain. Its valley sides gradually become more gentle. It flows south to meet the river Po at Pavia. Figure F shows the confluence. The river landforms are easy to interpret on this 1:200,000 (1cm:2km) map. The light green shading represents a nature reserve. The highly developed local infrastructure can be seen but there is no map evidence of the intensive farming. The river Po continues to flow east where it becomes larger, receiving more tributaries. It reaches the Adriatic Sea as a delta (see Figure A).

## ▼ Questions

**2** Draw a sketch map of the rivers Anza and Toce from the map extract and label as many river landforms as you can.

**3** What evidence is there that the river Toce has reached a lower section than the river Anza?

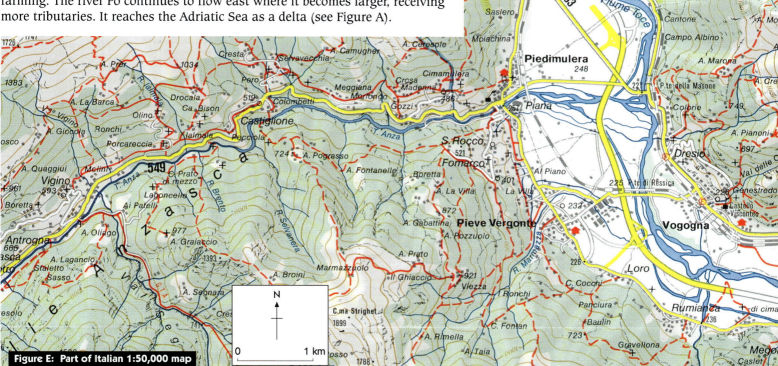

**Figure E: Part of Italian 1:50,000 map**

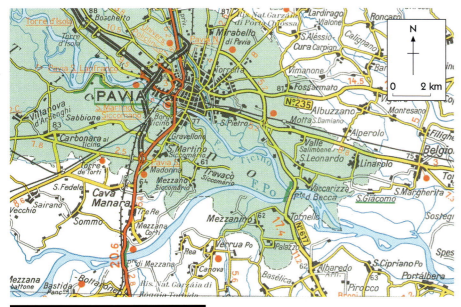

**Figure F: Part of Italian 1:200,000 map**

## ▼ Questions

**4** Describe the river landforms shown on Figure F.

**5** Draw some diagrams of the meanders on the Ticino to the west of Pavia to show how they may eventually develop into ox-bow lakes.

**Extension question**

**6** Describe and explain how the rivers you have studied in this Unit start in their upper sections with downward erosion and little deposition and end with sideways erosion and a lot of deposition. Quote the names of the rivers and locations in your answer. ➡

### Review

The rivers of the Po basin show a range of river landforms, from steep-sided glacial valleys to a wide flat plain across which the Po meanders.

# Human uses of rivers

The diagram (Figure A) shows a river drainage basin divided into an upper, middle and lower section to show the uses of a river. You will be able to add more to the diagram and perhaps more major categories of river use.

The uses of the river and its surrounding land depend on physical factors such as height above sea level, slopes, total precipitation and its regime, type of soil and underlying rock, and the quality of water. Human and economic factors are also important. These include the wealth of the region, the government policies towards farming and industry and the demand for leisure and tourism.

## ▼ Questions

1 Draw up three columns labelled Upper section, Middle section and Lower section. List the uses of the river shown in Figure A in the correct column. Now add further uses of rivers that you know of in each column.

2 Give some examples of how the natural landforms and shape of the river in Figure A have been changed.

3 Suggest a different use for each of the three sections of the Afon Nant Peris river in Snowdonia (page 58).

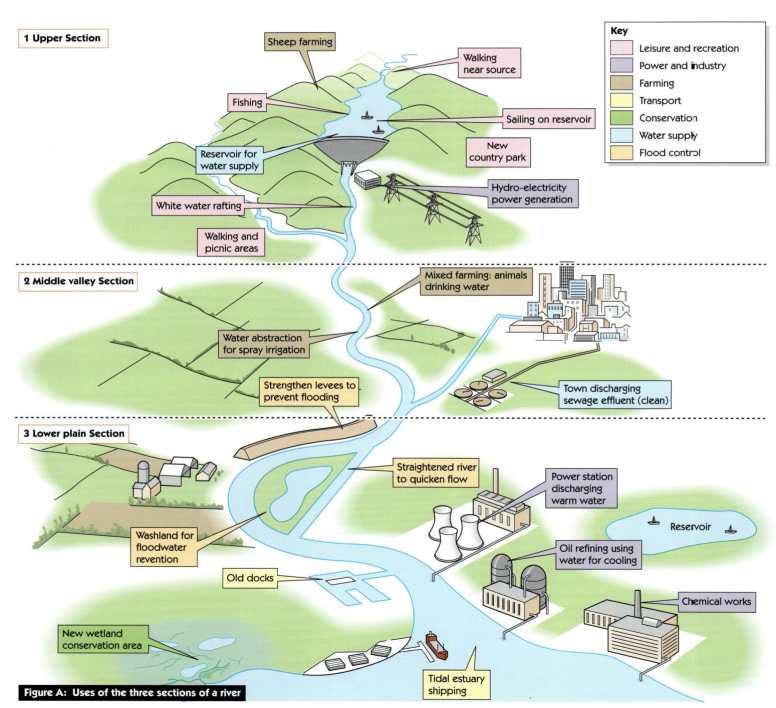

**1 Upper Section**

Sheep farming

Walking near source

Fishing

Sailing on reservoir

Reservoir for water supply

New country park

White water rafting

Hydro-electricity power generation

Walking and picnic areas

**Key**
- Leisure and recreation
- Power and industry
- Farming
- Transport
- Conservation
- Water supply
- Flood control

**2 Middle valley Section**

Mixed farming: animals drinking water

Water abstraction for spray irrigation

Strengthen levees to prevent flooding

Town discharging sewage effluent (clean)

**3 Lower plain Section**

Straightened river to quicken flow

Power station discharging warm water

Reservoir

Oil refining using water for cooling

Washland for floodwater revention

Old docks

Chemical works

New wetland conservation area

Tidal estuary shipping

**Figure A: Uses of the three sections of a river**

## Key ideas

- Rivers flood after heavy rain.
- Human interference with river courses can lead to problems.

## Key questions

1 What caused the serious 1997 floods?
2 Why was there much damage?
3 How can the risk of future flooding be reduced?

## Do you know?

? Floodplains: the natural areas over which rivers flood.
? Dyke: an embankment alongside a river built to prevent flooding.
? Surface run-off: the flow of water on the surface of the land.
? Interception: the 'holding up' of water movement by vegetation.
? Canalising: the process of straightening rivers to control the flow.

**Figure A: German soldiers move sandbags to shore up a dyke protecting Frankfurt-on-Oder from the rising floodwaters, as thousands were evacuated**

In July 1997 there was very heavy rainfall over Poland, the Czech Republic and Germany. In places rivers rose over five metres. More than 100 people lost their lives and the damage caused cost over £1.25 billion to repair. Over 600,000 hectares of land were flooded in south-west Poland. In the Czech Republic, one-third of all the land was under water. In Germany the land along the rivers Oder and Neisse was devastated and farms and villages were abandoned.

Like thousands of others along the banks of the Oder river, Mrs Brettner, aged 66, was driven out of her home last week by floods. The 130-year-old farmhouse she has shared with her husband Günter since 1952 is directly behind a dam that broke under pressure from the rising water.

The couple left at the last minute, packing a few necessities into the back of their old Trabant before the ground floor of the house was submerged.

"Valuables, documents, the photo album and the best things we have, like my husband's suit, and clothes for a couple of days. If they don't last, we'll just have to wash them," she said.

She managed to place her 22 geese with a farmer on a hill for safekeeping, but she was worried she may not have put out enough food for the hens she left on the roof of the house. There was no time to move the furniture to an upper floor, so the Brettners expect to return to a dismal sight whenever the floods abate.

Mrs Brettner has lived all her life by the Oder, crossing it by ferry to go to school as a child. She has seen big floods before but she claims that this experience has changed her view of the river for ever.

"It was never as bad as this. The water is flowing so quickly. I find the river frightening, threatening," she said.

The couple are sleeping on a sofa in their son's house near Wiesenau, a village that is almost entirely flooded. The little village church offered no services yesterday, its doors barricaded with sandbags as the water lapped around its whitewashed walls.

Wiesenau was abandoned by the emergency services on Friday after a nearby dyke collapsed and all inhabitants were ordered to leave their homes. The village was like a waterlogged ghost town yesterday, completely deserted except for a few policemen and rescue workers dozing in the sun and one or two stubborn residents who refused to leave.

**Figure B: From the Guardian, 28 July 1997**

The land along the west bank of the Oder is Berlin's 'vegetable garden'. Farmers in this flooded area owned 38,000 animals and 400,000 poultry. Local industry was ruined and chemicals, oil, fertilisers and sewage were washed into the floodwaters. Many people suffered health problems such as infectious rashes and diarrhoea.

During the clean-up programme, the army helped by removing the dead bodies of animals and using chemical disinfectants. Medical aid included vaccinations for tetanus, typhoid and hepatitis. European countries immediately sent aid such as water purification equipment, pontoon bridges and water pumps to the disaster areas

The newspaper article (Figure B) quotes Mrs Brettner as saying, 'it was never as bad as this.' She may well have been right. There are reasons why recent European floods have been very serious.

● Rivers such as the Neisse and Oder have been straightened and dykes built. Rivers flow faster and floodwaters move along more quickly.
● In eastern Europe the dykes may not have been maintained properly.
● Natural floodplains are not properly used as dykes stop rivers flooding into them.
● Surface run-off from towns has increased as more tarmac and concrete has been laid down.
● The farmland does not absorb so much water because it has been compacted by heavy machinery. Fertilisers also damage soil structure so it does not absorb water well.
● Trees and vegetation have been cut down so there is less interception of water.

## Preventing future floods

From Figure C you will see that German and Polish rivers start in the Czech Republic: they are international rivers. Because of this, international co-operation is needed to manage river flow. If potential floodwaters can be held back in the upper sections of the drainage basins, then there is more time to warn and if necessary evacuate people who live in the lower sections.

Some of the possible methods for preventing floods are shown in Figure D. Many of the ideas reverse the older methods of flood control where dykes were built. The newer ideas try to reinstate the river channel and floodplain as the 'natural way' to deal with floods. Reducing surface run-off by building fewer car parks, factories and roads may be a difficult solution for the developing economies in parts of Europe.

### ▼ Questions

1 Write brief notes on the floods in eastern Europe using the following headings: When? Where? What? Who? Why?
2 Why do you think Mrs Brettner might be right when she said, 'it was never as bad as this.'?
3 List some different types of assistance and aid that were provided to help the flooded areas.
4 Why is international co-operation likely to be important when trying to prevent future floods?
5 How will the following methods help to prevent flooding along the Oder south of Berlin?
   a  Planting trees in the upper Oder and Neisse rivers
   b  Maintaining dykes along the Oder
   c  Building a flood relief channel and flood retention area
   d  Stopping plans to straighten the Oder
   e  Moving farmhouses from the floodplain and taking some farmland out of use.
6 Floods are often seen as a natural hazard. From what you read, do you consider them as a natural or a people-made hazard?

**Figure C: Location of the 1997 Eastern Europe floods**

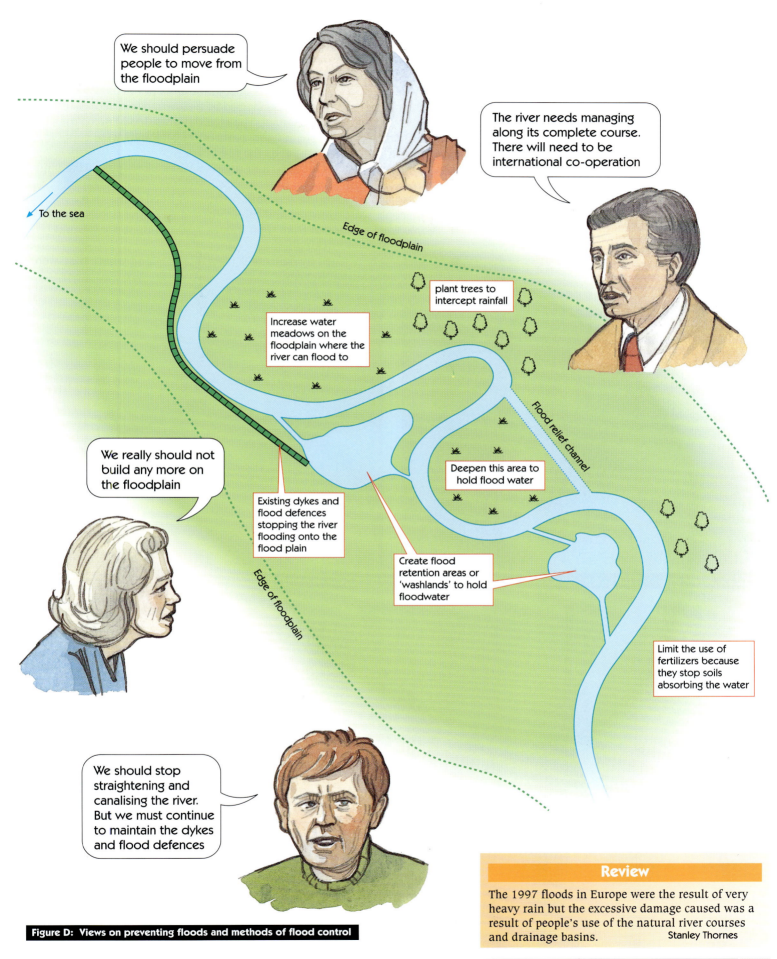

Figure D: Views on preventing floods and methods of flood control

## Key ideas

● Floods occur after heavy rain but are becoming worse because of poor drainage systems and the increase of buildings on floodplains.

## Key questions

● What caused the 1998 floods?
● Why did people have so little warning?

In the Midland counties of England rain fell all day during Thursday, 9 April 1998. Parts of the region received as much rain in 24 hours as the monthly average for April. The rainfall was caused by a slow-moving depression (low pressure) over the English Channel. The next day was Good Friday, the start of the Easter holiday. People in many parts of the Midlands were going to have to cope with the worst floods for 150 years.

● Major roads such as the M40 and the A46 Warwick by-pass were shut for several hours.
● Banbury railway station was shut because of floodwater.
● Two people were drowned and three others died in the floods.
● People were evacuated from their homes – 200 in Banbury, several hundred around Buckingham.
● Anglers were rescued by helicopter from a lake island where they were fishing near Milton Keynes.

**Figure A:** Before and after: the bridge at the Pump Room gardens when the River Leam burst its banks

**Figure B:** Before and after: a main junction in Leamington Spa

- Pensioners were rescued from a caravan site.
- Flood damage was estimated at £1.5 billion (higher than the bill for damage from the 1987 hurricane).
- After the flood people's insurance premiums were raised.

## Leamington Spa

On Good Friday the River Leam broke its banks at 5.00 a.m. By 7.00 a.m. the Pump Room Gardens in the town were under a metre (3 feet) of water (Figure A). People woke up to find their back gardens and cellars under water. Streets were flooded and shops could not open (Figure B). Firefighters trying to reach people and properties had to hang onto lamp-posts to stop themselves being washed away down the street.

The floods in the town were the worst in living memory. Damage to property was considerable. People were told that they could not pump out their homes because there was nowhere for them to pump the water to. Shops had much of their stock damaged (Figure C).

Local residents had no warnings of the floods. The reasons for this are that the floods had risen very quickly during the night and floods are not very common in the town. The reasons why the floods had risen so quickly are more complex.

Six million people now live on floodplains in the UK. Rivers have been straightened and canalised to make water flow quickly. In wet weather the flow of water should be slowed down by meandering rivers crossing floodplains. Once the rivers reach the top of their banks (**bankfull**) the water should flow onto the floodplains to be absorbed.

Figure D shows a small river and its floodplain in the Leamington area; any future building on this land would not be advisable. The Environment Agency has issued new guidelines to planning authorities to halt building on floodplains. In many places the new guidelines are too late.

Figure D on page 65 shows some of the methods suggested for preventing floods in Europe. The methods are the same for preventing floods in the UK and elsewhere in the world.

Figure C: The aftermath of the flooding

Figure D: A small river and its floodplain near Leamington Spa

## ▼ Questions

1 What were the reasons for the floods in the English Midlands? Divide your answer into Natural causes and People-made causes.

2 You are a lifelong resident of Leamington Spa. Write a letter to the local council expressing concern over the lack of warning you had of the floods in Easter 1998. Also write of your concerns about future flooding and possible future building on the local floodplains.

### Do you know?

? Flood defences built around Maidenhead, Windsor and Eton cost £80 million, or an average of £14,000 for every new house.

? By 2016 the number of people living on floodplains will increase by the equivalent of a city the size of Bristol, so increasing the risks of more major floods.

? It is more cost effective to work with nature rather than against it.

### Review

The floods of Easter 1998 had both natural and people-made causes. Floodplains and natural river courses will have to be better preserved to absorb floodwaters.

# 7

# The coastal system

## Main activity

Describing, explaining and interpreting. Following through a fieldwork exercise.

## Do you know?

? The sea erodes the land in four ways easily remembered by the word CASH.
**Corrasion:** the waves throw pebbles and rocks against the cliffs, wearing them away.
**Attrition:** the pebbles and rocks are worn away as they crash against each other in the water.
**Solution:** the water itself is slightly acidic and can dissolve minerals such as calcium carbonate in chalk and limestone.
**Hydraulic action:** the waves can exert a pressure of up to 30 tonnes per square metre. Air is compressed into cracks and joints in the rocks. When the water retreats the compressed air is released with explosive energy loosening rock particles.
? The sea transports eroded material by longshore drift.
? Deposition can be in the form of a beach, bar, spit or tombolo.
? When sea level changes, coastal landforms such as raised beaches, rias and fjords may result.

To study this unit on coasts you should already know enough about coastal processes and landform features to interpret case studies.

Figure A explains the coastal system. The labelled photographs, Figures B to D describe and explain the formation of several examples of coasts. Figure D shows a range of human uses. In the UK and many parts of Europe it is difficult to analyse a coast from only a physical viewpoint.

## The coastal system

The coastline can be viewed as a system with inputs and outputs. The inputs are sand, mud, gravel and other eroded materials forming the beach. The sea, through the process of longshore drift, transports this material along the coast until it is deposited somewhere else. Provided the amount of material coming into the coastal zone equals the amount of material leaving it, the section of coast is said to be in balance or in equilibrium. This equilibrium can be upset by various natural events such as rising or falling sea levels; however, nowadays the coastal equilibrium is usually upset by the actions of people, such as by the construction of groynes and other coastal defence works.

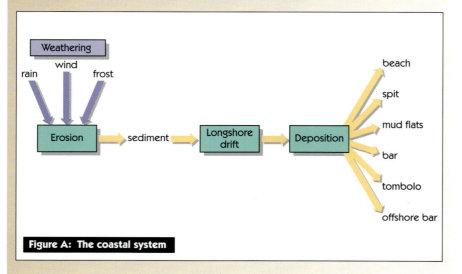

Weathering — rain, wind, frost → Erosion → sediment → Longshore drift → Deposition → beach, spit, mud flats, bar, tombolo, offshore bar

**Figure A: The coastal system**

near vertical chalk cliff

wave-cut platform

horizontal bands of hard flint

evidence of waves smoothing and indenting base of cliff

beach of chalk and flint pebbles

sand beach

**Figure B: Chalk cliff at Birling Gap, East Sussex**

## Coastal erosion: Marsden Rock

Figure E shows a famous feature of the Tyne and Wear coastline, Marsden Rock, in 1992. This natural **arch** had been a popular tourist attraction for over a century. The same site in 1997 is shown in Figure F. The natural arch collapsed into the North Sea in the spring of 1996, leaving a heap of limestone rubble and two smaller Marsden rocks which are examples of **stacks**. Warning signs that the arch was about to collapse came during a cold spell in early February 1996 when frost action widened existing cracks and triggered rockfalls.

At one time Marsden Rocks were joined to the mainland. Erosion created a series of landforms. The collapse of the arch is just one stage in the gradual retreat of the coastline.

The rate of erosion largely depends upon the type of rock forming the coast. The limestone of Marsden Rocks is fairly resistant to erosion, as is the chalk cliff shown in Figure B. However, the rock forming the Holderness coast of East Yorkshire (see Mappleton pages 71–73) is weak boulder clay which is easily eroded.

coniferous trees planted to stabilise sand dunes

sand dune eroding with high tides and wind

sea edge

marram grass only slightly stabilising sand dunes

active sea erosion at high tide

beach

**Figure C: Eroding sand dunes at Culbin Sands, Moray**

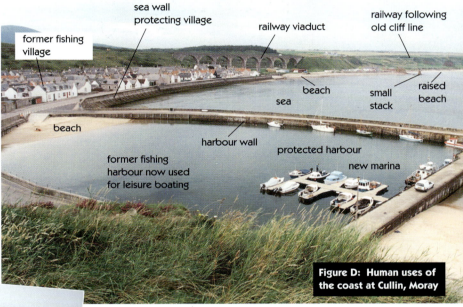

sea wall protecting village

railway viaduct

railway following old cliff line

former fishing village

beach

small stack

raised beach

sea

beach

harbour wall

former fishing harbour now used for leisure boating

protected harbour

new marina

**Figure D: Human uses of the coast at Cullin, Moray**

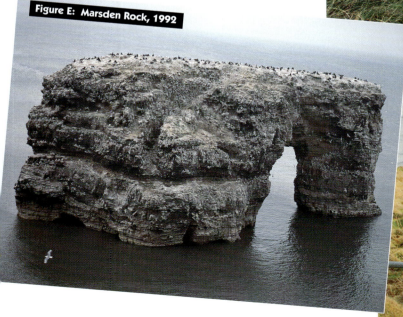

**Figure E: Marsden Rock, 1992**

**Figure F: Marsden Rocks, 1997**

## Coastal deposition: Spurn Head

Figure G is an aerial view of Spurn Head. This is a 6 kilometre ridge of sand and shingle extending nearly halfway across the mouth of the Humber estuary. Spurn Head is an example of a feature called a **spit**. The spit forms a sweeping curve which has been built up by longshore drift from the north. It is here that some of the material from the cliffs at Mappleton is deposited.

Although Spurn Head is a feature of deposition it is also being eroded, partly because it is being starved of beach material. This is because of the coastal defences further up the coast at places such as Hornsea, Mappleton and Withernsea. The spit has been destroyed four times during the last few centuries (Figure H). In 1997 the sea breached the spit, damaging the road and cutting off the small settlement at the tip of the spit, Spurn Point.

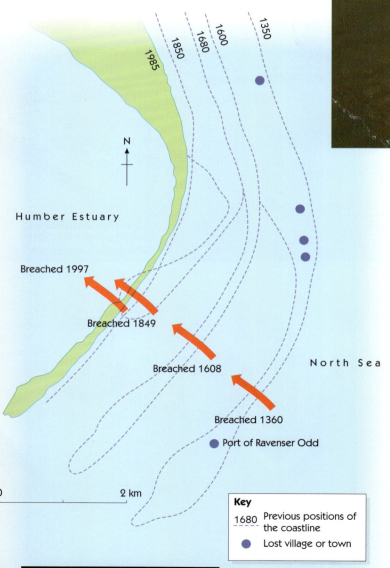

**Figure H: The movement of Spurn Head**

Key

1680 — — Previous positions of the coastline

● Lost village or town

Figure G labels: longshore drift · usual wind and wave direction · small settlement · sand and mud · sheltered side of spit · road · deposits build up · lighthouse · lifeboat station · second most usual wind and wave direction · River Humber · sand dunes colonised by marram grass

**Figure G: Spurn Head, aerial view**

### ▼ Questions

1 a How does the sea erode the coast?
  b How does the sea transport and deposit material? You will need to refer to the Do you know? box.

2 Describe and explain the coastal features shown in either Figure B or C.

3 In what ways have people used and changed the coast shown in Figure D?

4 Draw sketches to show how Marsden Rock became Marsden Rocks. On your sketches annotate the following details:
  a rock type
  b evidence that some parts are more resistant to erosion than other parts
  c evidence of active sea erosion
  d whether the tide is high or low
  e names of the landform features.

5 a Describe the scene in Figure G.
  b Explain how this spit was formed.
  c In what ways has this spit been destroyed over the past few centuries?
  d From the dates given is it true to say that there is a cycle of spit breaching?
  e Why might beach starvation be the cause of the breaching?

## 'If it goes on like this, there won't be any farms left'

Sue Earle lives on what is left of Grange Farm at Cowden, near Hornsea in East Yorkshire. Over the past eight years, she has lost eight acres to the sea, a third of her arable land, writes Rosemary Behan.  In November 1996, her three-bedroom Victorian farmhouse was demolished by East Yorkshire council because it was just five yards from the cliff edge and considered too dangerous to live in. The same thing happened to several of her neighbours.

Ms Earle and her uncle, David, who lives with her, received a bill for £3,500 that they have still not paid.

They now live in a wooden hut they built at their own expense and are suing the council for compensation. On hearing yesterday's announcement by the agriculture select committee that more farmland should be allowed to fall into the sea, Ms Earle, 45, said: 'They want shooting. Let them come and live in my hut, and I'll go to their houses. Why did we fight two world wars? To protect our land. They're taking every inch of land to build houses on. If it goes on like this, there won't be any agricultural land left.

'I don't think they should protect every inch of land, but they should try to save as much as possible. I lost my farm, which was worth £250,000, and my loss of income for all this time has been huge.'

Ms Earle says erosion of her farmland accelerated after sea defences were built to protect the village of Mappleton, one mile north of Cowden, in 1991. Two rock groynes were built and sediment was prevented from drifting along the coast to the beach.

'The cliff erodes faster because there is no beach to protect it,' she said.

**Figure A: from the Daily Telegraph, 6 August, 1998**

The article in Figure A was written in 1998. Similar articles have appeared in local and national newspapers throughout the 1990s. People living near the sea naturally want their land and property saved from coastal erosion. It has been policy to protect the coast from erosion and flooding, and sea defences of many types have been built during the 20th century. This policy will most likely be changed during the 21st century.

The article in Figure B suggests that tracts of land along the east and south-east coasts of England should be surrendered to the sea. The Commons Agriculture Committee report said, 'It is time to declare an end to the centuries-old war with the sea and seek a peaceful accommodation with our former enemy.' Peter Luff, the committee chairman, said 'We must work with nature and not against it. We must be a little more humble about our relationship with it.'

The fieldwork for this unit was carried out at the same time as the two articles were written. It looks at an area where hard engineering methods have been used to protect a stretch of coastline. Some softer approaches are being considered for the future. In areas such as Cowden (Figure A), a managed retreat of the coastline might be the answer, with people like Sue Earle receiving compensation for moving and losing land.

## It is time to admit defeat on coastal defence, say MPs

*Paul Brown*
*Environment Correspondent*

Parts of the British coast at risk of flooding should be abandoned to the sea, a committee of MPs suggested yesterday.

Continuing to build ever higher defences to keep out the rising sea is no longer an option, and retreat to new positions inland should begin immediately in some places, the Commons Agriculture Committee said yesterday.

People who are thus forced to abandon homes and fields for the general good of the community should be compensated by central government, the MPs say.

**Figure B: From the Guardian, 6 August, 1998**

CLIFF EROSION KEEP AWAY FROM EDGE

## Investigation at Mappleton

### Aims

To investigate the effects of the sea defences at Mappleton, which is a village to the south of Hornsea in East Yorkshire.

To see if the sea defences (built 1991) had in fact allowed beach material to build up, thus starving the coast to the south of beach material.

### Methods

To use primary and secondary sources. The primary methods (fieldwork) would be taking beach measurements, sketching and taking photographs.

The secondary sources used would be OS maps, newspaper articles and a Coastal Data Pack produced by the local council.

### Results

The results are presented within this unit as Figures C to I.

**Figure C: Geology of the Holderness peninsula**

Flamborough Head
Boulder clay overlying chalk cliffs
Bridlington
Great Driffield
Hornsea
Mappleton
Beverley
Hull
North Sea
Withernsea
River Humber
Spurn Head

Key
Alluvium
Boulder clay
Chalk

0    10 km

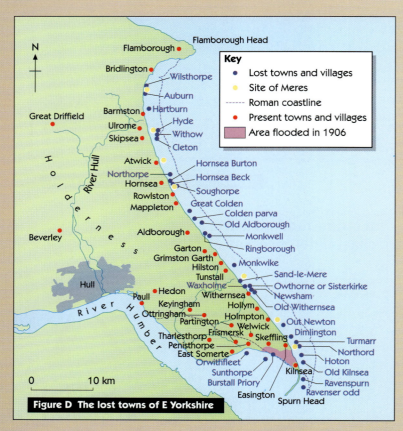

**Figure D  The lost towns of E Yorkshire**

Key
- Lost towns and villages
- Site of Meres
- Roman coastline
- Present towns and villages
- Area flooded in 1906

N

Flamborough Head
Flamborough
Bridlington
Wilsthorpe
Auburn
Great Driffield
Barmston · Hartburn
Hyde
Ulrome · Withow
Skipsea · Cleton
Atwick · Hornsea Burton
Northorpe · Hornsea Beck
Hornsea · Soughorpe
Rowlston · Great Colden
Mappleton · Colden parva
Old Aldborough
Aldborough · Monkwell
Ringborough
Beverley
Garton · Monkwike
Grimston Garth
Hilston · Sand-le-Mere
Tunstall · Owthorne or Sisterkirke
Waxholme · Newsham
Hedon · Withernsea · Old Withernsea
Paull · Keyingham · Hollym
Ottringham · Holmpton · Out Newton
Partington · Welwick · Dimlington
Frismersk · Turmarr
Tharlesthorp · Skeffling · Northord
Penisthorpe · Hoton
East Somerte · Old Kilnsea
Orwithfleet · Kilnsea · Ravenspurn
Sunthorpe · Ravenser odd
Burstall Priory
Easington · Spurn Head

Holderness
River Hull
Hull
River Humber

0   10 km

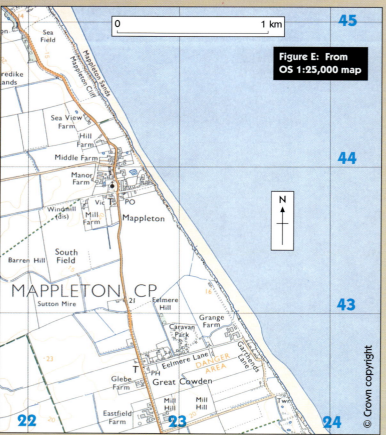

**Figure E:  From OS 1:25,000 map**

© Crown copyright

0   1 km

45

Sea Field
Mappleton Sands
redike lands
Sea View Farm
Hill Farm
Mappleton Cliff
Middle Farm

44

Manor Farm
Windmill (dis) · Mill Farm
Vic · PO
Mappleton
South Field
Barren Hill

43

MAPPLETON CP
Sutton Mire
Eelmere Hill
Grange Farm
Caravan Park
Eelmere Lane
DANGER AREA
Garthends Lane
Glebe Farm · PH
Great Cowden
T
Eastfield Farm
Mill Hill · Mill Hill

22          23          24

N

**Figure F:  Beach profiles at four points**

Beach profile 10m north of the northern groyne

Break at 8m

Cliff base open to wave check

Sea
0   10   20   30   40   50   60   metres

Beach profile 10m north of the southern groyne

Granite boulders

to cliff base protected from wave attack

Sea
0   10   20   30   40   50   60   metres

Beach profile 10m south of the southern groyne

This beach measured at 1.9m lower than north

Granite boulders

Sea
0   10   20   30   40   50   60   70   metres

Narrower beach with cliff foot at the same height as protective boulders at Mappleton

Beach profile 400m south towards Cowden

Base of cliff exposed to sea at high tide (Figure I)

Sea
0   10   20   30   40   50   metres

**Figure H:** The main boulder groyne at Mappleton showing the lower beach on the right (south)

**Figure I:** The easily eroded cliffs south of Mappleton towards Cowden. Here the beach is narrow and the sea reaches the cliffs at high tide

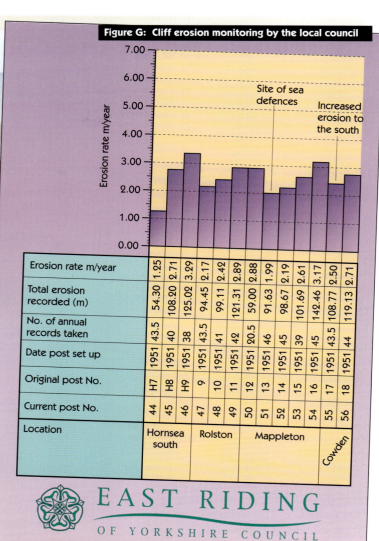

**Figure G: Cliff erosion monitoring by the local council**

| Erosion rate m/year | 1.25 | 2.71 | 3.29 | 2.17 | 2.42 | 2.89 | 2.88 | 1.99 | 2.19 | 2.61 | 3.17 | 2.50 | 2.71 |
|---|---|---|---|---|---|---|---|---|---|---|---|---|---|
| Total erosion recorded (m) | 54.30 | 108.20 | 125.02 | 94.45 | 99.11 | 121.31 | 59.00 | 91.63 | 98.67 | 101.69 | 142.46 | 108.77 | 119.13 |
| No. of annual records taken | 43.5 | 40 | 38 | 43.5 | 41 | 42 | 20.5 | 46 | 45 | 39 | 45 | 43.5 | 44 |
| Date post set up | 1951 | 1951 | 1951 | 1951 | 1951 | 1951 | 1951 | 1951 | 1951 | 1951 | 1951 | 1951 | 1951 |
| Original post No. | H7 | H8 | H9 | 9 | 10 | 11 | 12 | 13 | 14 | 15 | 16 | 17 | 18 |
| Current post No. | 44 | 45 | 46 | 47 | 48 | 49 | 50 | 51 | 52 | 53 | 54 | 55 | 56 |
| Location | Hornsea south | | | Rolston | | | Mappleton | | | | | | Cowden |

EAST RIDING
OF YORKSHIRE COUNCIL

## Analysis

Little has been written on this as you should complete it yourself.

## Conclusions

The sea defences at Mappleton have been successful in trapping beach material. There has been little or no erosion at the point of the defences; the road through Mappleton seems safe. There is evidence of faster erosion to the south of the defences. Sue Earle's land is disappearing quicker than it used to. In the future government policy may be to let the land erode and abandon the sea defences.

## Further research

Over the next few years the area needs to be re-visited and evidence collected. The actions resulting from the 1998 Commons Agricultural Committee should be studied. Other sea erosion areas should be studied to look for similarities and differences.

## ▼ Questions

1  a  Where is Mappleton?
   b  Why was Mappleton being eroded so quickly?
2  How many towns and villages have been lost to the sea since Roman times? (See Figure D.)
3  What hard engineering works were carried out at Mappleton?
4  What fieldwork methods were used to see if the groynes were collecting beach material?
5  a  What does Figure H show to indicate that the Mappleton defences do not help the coastline to the south?
   b  How does this evidence link with information in Figure A?
6  Write a short Analysis and Conclusion to the fieldwork using the results presented in Figures C to I.

## Review

The coastal system includes processes and landforms of erosion and deposition. People have interfered with the system to protect and to gain land. Mappleton may have benefited from hard engineering methods but other places have suffered because of them. Soft approaches including managed retreat may be the future government policy towards fragile coastlines.

# The human use of the coast

## Main activity

Interpreting satellite images and linking them with atlas maps. Writing reports.

## Key questions

● What is the Black Sea and its surrounding coastlines like?
● What problems does such an enclosed sea have in the modern world?
● What are the coastal landscapes of the Severn Estuary?
● How do the complex ecosystems cope with the increasing development around the estuary?

Throughout history, people have used the coast with its beaches, inlets, natural harbours and land immediately behind for a variety of uses. Figure D, page 69 shows the coastline has potential for fishing, settlement, transport and tourism. Offshore there are oil and gas rigs; the port of Aberdeen in Scotland is an offshore service centre.

In many parts of the world the coasts are being threatened by urban and rural development. Careful management of coastlines, estuaries and coastal wetlands can result in **sustainable development**. This means there are close links between conservation and meeting the demand for leisure and other activities. In the UK there are a range of agencies and bodies that look after the coast e.g. English Nature, the Countryside Commission and the National Trust. In France the Rhône delta called the Camargue is protected by Camargue Nature Park.

In the next three pages the physical and human geography of two sea areas are studied. The method is the enquiry approach, outlined in the Key questions. In both cases secondary sources are used. You can follow up more ideas and issues using the Internet, CD-ROMs and books.

## Do you know?

? Delta: where a river enters the sea through a system of distributaries. Deltas form where there are no great sea currents such as in the Mediterranean and the Black Sea.
? Wetland: a nearly level area with water-logged surfaces and high water tables. Often drained for farming thus ruining a natural wildlife habitat.
? Fold mountains: mountain ranges formed by the movements of crustal plates.

## CASE STUDY: The Black Sea

The Black Sea has no link to an ocean except indirectly through the Bosporus to the Mediterranean Sea. It has a surface area five times smaller than its catchment area which contains 162 million people. It receives water from many large rivers such as the Danube, Dniester and Dnieper (see Figure A, and an atlas). The salinity (salt) level of the Black Sea is half that found in an ocean. The climate is very cold in winter as there are prevailing northerly winds; severe storms also occur. The Black Sea is, however, the warmest sea area around the whole of Russia.

Geologically the Black Sea is a basin, with a greatest depth of 2135 metres. The Sea is bordered to the west, south and east by young fold mountains which experience earthquakes. The Pontine mountains in north-eastern Turkey are high, and spectacular cliffs rise out of the sea in the Trabzon area (Figure B). The Caucasus mountains further north have peaks over 3500 metres high.

The Danube delta is the largest delta in Europe (Figure C). Through this delta flow millions of tonnes of dissolved and suspended substances. Pollution enters the sea from other rivers and ports making the Black Sea an environmental disaster area. Figure D outlines some of the problems of the Sea.

There are attempts to save the Black Sea. Scientists from Black Sea states have taken legal action against Austria and Germany to force them to stop polluting the river Danube. In 1996 the governments of the Black Sea states agreed a 'Strategic Action Plan for the Rehabilitation and the Protection of the Black Sea'. The plan is to design conservation areas which have regional significance. The states realised that protecting the Black Sea needed integrated policies which covered more than one country.

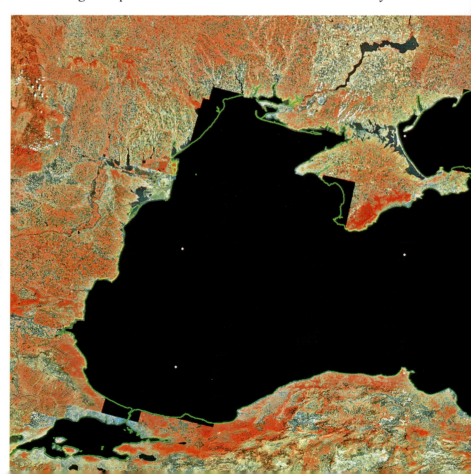

Figure B: Caucasus mountains bordering the eastern Black Sea

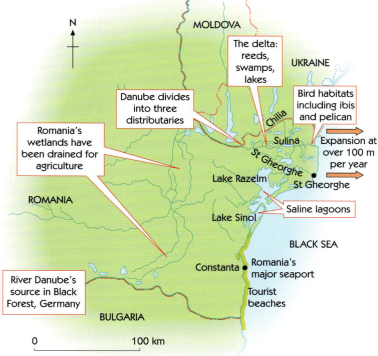

The delta: reeds, swamps, lakes

Danube divides into three distributaries

Bird habitats including ibis and pelican

Romania's wetlands have been drained for agriculture

Expansion at over 100 m per year

Saline lagoons

Romania's major seaport

Tourist beaches

River Danube's source in Black Forest, Germany

MOLDOVA

UKRAINE

Chilia

Sulina

St Gheorghe

Lake Razelm

St Gheorghe

ROMANIA

Lake Sinol

BLACK SEA

Constanta

BULGARIA

0          100 km

**Figure C: The Danube delta region of the Black Sea in Romania**

Sewage treatment works in Odessa, Ukraine broke down causing outbreaks of cholera in 1996

Oil is pumped to the Black Sea port of Novorossiysk from the Caspian Sea, Kazakhstan and Russia. Fertilisers wash into the sea. 750,000 tonnes of nitrogen a year cause algae blooms

Danube flows through 11 countries dumping its polluted load in Romania

Busy shipping lanes and oil spillages

'The Black Sea is very severely damaged, but it's not yet dead.'
Laurence Mee, environmentalist

Fish stocks have declined; 40 species now extinct

Tourist resorts in decline e.g. around the Crimea

Pesticides from intensive farming wash into the sea

Dolphins seriously threatened

Chemicals and industrial waste from Russia and European countries

**Figure D: The problems of the Black Sea**

**Figure A: Satellite image of the Black Sea**

## ▼ Questions

1 Closely compare the satellite map, Figure A, with an atlas map. Draw a sketch map of the Black Sea area labelling the main rivers, mountains and countries.

2 Write a format for an enquiry with the **Aim** of answering: What makes the Black Sea unique? What problems threaten the Sea and surrounding coastlines? What can be done to save the Black Sea environment?

3 Complete the enquiry briefly using the resources presented in this unit and others you can find.

4 Why do you think it will be difficult to carry out an integrated conservation programme for the Black Sea region?

## Review

The Black Sea is a deep basin surrounded by a variety of coastlines. It has been polluted to dangerous levels and there are now strategies to save it.

The river Severn is the longest river in the UK and it ends in the Severn Estuary which leads into the Bristol Channel and then the Celtic Sea. The bed of the estuary is bare rock or shifting banks of sand and gravel. The area has the second largest **tidal range** in the world (see Do you know? box). The famous Severn bore (Figure D) can reach 3 metres high. It is highest at spring tides and especially during the times of the spring and autumn equinox tides. It can be even higher if these tides coincide with the prevailing south-westerly winds and a **storm surge**.

The estuary is an area of wilderness running through the middle of a heavy industrialised and settled area (see Figure A and an atlas). Figure B gives details about some of the features of the Severn Estuary. Figure C shows some of the threats to its habitats. Sea level rise is a long term threat.

- Over 11 species of fish: an area of ocean fish breedin (a nursery)
- Migratory fish such as the salmon and eel
- Water quality has recently improved with more sea trout near the Usk estuary
- A reef-building honeycomb worm builds up reefs which are colonised by other species
- Tens of thousands of migratory birds including geese
- Salt marsh on the higher shores: a large share of Britain's total
- Water transport never developed high up the estuary although Gloucester was a small port
- The first Severn road crossing opened 1966
- The second Severn crossing opened 1996

**Figure B: Features of the Severn Estuary**

**Figure A: Satellite image of the Severn Estuary**

Figure D: Surfing the Severn bore

## Protecting the estuary

The Wildfowl Trust manages an area of land along the Severn at Slimbridge which is well known for its winter geese and Bewick swan migrants. The National Trust owns two headlands near Weston-super-Mare. There is a Nature Reserve in the bay near Bridgwater. In 1995 the estuary was designated a Special Protection Area but the protection only reaches as far as the low tide mark. The offshore reefs are not yet protected but they could be under the European Habitats Directive.

There is no overall integrated plan to manage and protect the estuary or to regulate development. The conservation measures seem to be piecemeal, as do the developments taking place all around the estuary. This is no different to situations in other fragile areas in the UK and Europe.

Wessex Water, the local water company, has taken the first steps to recreate a salt marsh habitat at its new Weston-super-Mare sewage plant. This is located at Bleadon Levels to the south of the town. The plan will undo centuries of land reclamation. Sea walls will be breached and a new seawall will be built further inland. The area behind the breach will become a new salt marsh and a habitat for birds. This is called **managed retreat** and is in keeping with new proposals to let the sea take over land it should naturally cover (see the government proposals page 71). A footpath and cycle way will form part of the Avon Tidal Trail.

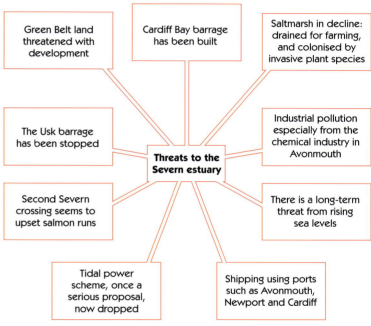

Green Belt land threatened with development

Cardiff Bay barrage has been built

Saltmarsh in decline: drained for farming, and colonised by invasive plant species

The Usk barrage has been stopped

Threats to the Severn estuary

Industrial pollution especially from the chemical industry in Avonmouth

Second Severn crossing seems to upset salmon runs

There is a long-term threat from rising sea levels

Tidal power scheme, once a serious proposal, now dropped

Shipping using ports such as Avonmouth, Newport and Cardiff

**Figure C: Threats to the Severn estuary.**

### Do you know?

**?** Tidal range: the vertical distance between tidal low water and high water. The Bay of Fundy in eastern Canada has the world's largest tidal range at 14–15 metres.

**?** High tide: occurs twice a day; it is 52 minutes later each day.

**?** Spring tides: occur every 14 days when the gravitational attraction of the moon and sun are greatest.

**?** Neap tides: occur between the spring tides when the gravitational attraction of the moon and sun are least.

**?** Equinox: the times of the year when the sun is overhead at the equator.

**?** Storm surge: a change in sea level caused by extreme winds.

**?** Bore: a wave of water moving up a funnel-shaped estuary such as the Severn.

**?** Salt marsh: vegetated mudflats along low lying coastal areas.

## ▼ Questions

**1** Closely compare Figure A with an atlas map of the same area. Draw a sketch map of the area shown on the image. Mark in and label the rivers Usk and Avon. Label the main towns. Mark in the two Severn Bridge crossings. Label the Bleadon Levels immediately to the south of Weston-super-Mare.

**2** Write a report for a group of visiting European conservationists which outlines the natural features of the Severn estuary. Then make a list of possible threats to the estuary.

**3** In what ways would you like to see the Severn Estuary protected and conserved for the benefit of wildlife and local people?

### Review

The Severn Estuary has one of the world's highest tidal ranges. A wide range of uses has developed along its shores. There are threats to the estuary and some protective measures are now in place.

# 8 Highland glaciation

## Key ideas

● The landscape we see today in several of Britain's highland regions has been greatly affected by glacial action during the Ice Age.
● Glacial ice can erode, transport and deposit vast amounts of material.

### Main activity

A study of the main features of highland glaciation including the study of photographs of landforms.

### Do you know?

? Hippopotamuses, rhinoceroses and elephants lived in Britain 100,000 years ago.
? The Ice Age may not be over and the ice sheets may return to Britain.

## Ice today

The least hospitable place on Earth is the continent of Antarctica. This vast desert of ice suffers from extreme cold and strong winds throughout the year. Winter temperatures regularly fall below –50°C. Ice covers 10% of the Earth's land area and 90% of that ice is in Antarctica. In places the ice covering Antarctica is over 4 kilometres thick.

At the North Pole the Arctic Ocean is ice-covered, but there is much more ice in the ice sheet which covers Greenland. Up to 3 km thick, the Greenland ice sheet contains 8% of the world's ice. Smaller areas of ice called ice caps are found in Iceland and Norway.

## The Ice Age

At present the Earth has two ice sheets. In the past there were two others, one in North America and one in Europe. At that time almost a third of the planet's land area was covered by ice.

The Ice Age began about 3 million years ago; since then there have been over twenty advances and retreats of the ice sheets. Figure A shows the area covered by the maximum extent of the ice sheets, though for much of the Earth's history there have been no ice sheets at all. The last glaciers in Britain melted about 10,000 years ago. The Ice Age has not ended, however, since two ice sheets still exist.

During the Ice Age there have been warmer periods called interglacials when the ice retreated. It is probable that we are currently living in an interglacial period; some previous interglacials have had higher temperatures than at present and there was less ice around; fossils of lions, rhinoceroses and elephants have been found in England suggesting that the climate was warmer than it is now. It is known that the temperature of western Europe reached a

### ▼ Questions

1 a How long ago did the Ice Age begin?
  b How many times during the Ice Age have the ice sheets advanced and retreated?
  c What causes the ice sheets to advance and retreat?
2 a What is an interglacial period?
  b Why do most scientists think that we are now living in an interglacial period?

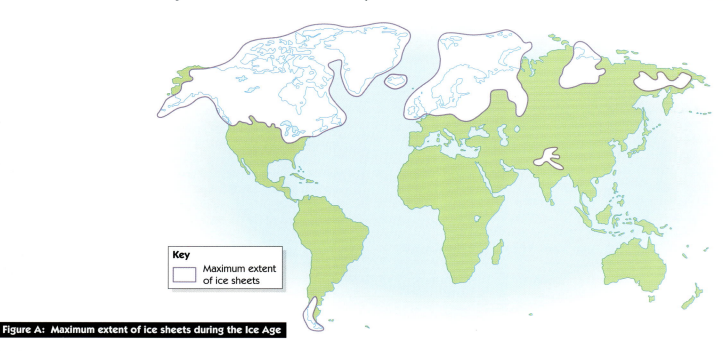

Key
☐ Maximum extent of ice sheets

**Figure A: Maximum extent of ice sheets during the Ice Age**

maximum about 7000 years ago (called the climatic optimum) and has decreased since then. This fall in temperature may soon be enough to cause another glacial advance. It is thought, for example, that the current Norwegian ice cap is not a remnant of the earlier Scandinavian ice sheet but a more recent development due to the fall in temperatures since the climatic optimum.

The Earth is a planet still gripped by the Ice Age and one day the ice sheets may return to Europe.

## Ice formation

Ice is formed when layers of snow build up, year by year, and do not melt during the summer. The lower layers are gradually compressed by the increasing weight of snow above. Air is forced out of the snow and it slowly turns to ice. Newly fallen snow has a density of less than 0.1 – that is, less than 10% snow and over 90% air. After just one winter the snow has compacted and its density increased to 0.5; such compacted snow is called firn or névé. True glacier ice, with a density of 0.9 or more, takes decades or even centuries to form.

## Glacier movement

Most glaciers have their sources in mountainside hollows called corries (also called cirques or cwms). Corries begin (see Figure B) as much smaller hollows called nivation hollows which fill with snow. Freeze-thaw action slowly increases the size of the hollow and the snow slowly turns into ice. As the ice in the nivation hollow increases in depth it begins to slip, gouging out the hollow and creating a corrie. Further increases in ice depth will cause the ice to flow over the lip of the corrie and move downhill under the influence of gravity and the weight of ice upstream.

Glacier movement is a complex process. The ice flows slowly downhill, its base sliding across the valley floor and its interior deforming, creeping and breaking. Within the glacier, rates of flow usually decrease towards the base because of friction with the valley floor.

Rates of glacier flow are often as little as a few centimetres a day. Rarely do rates of over a metre a day occur, even at the surface. Occasionally movement may be much quicker; 75 metres a day was achieved by Alaska's Black Rapids Glacier in 1937. Such rapid movements are called glacier surges and are probably caused by increased snowfall or earth tremors.

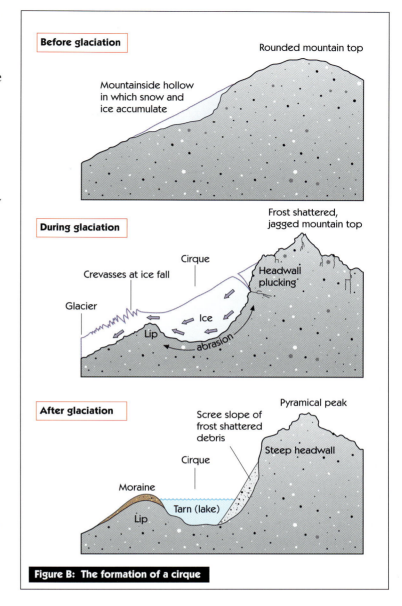

**Figure B: The formation of a cirque**

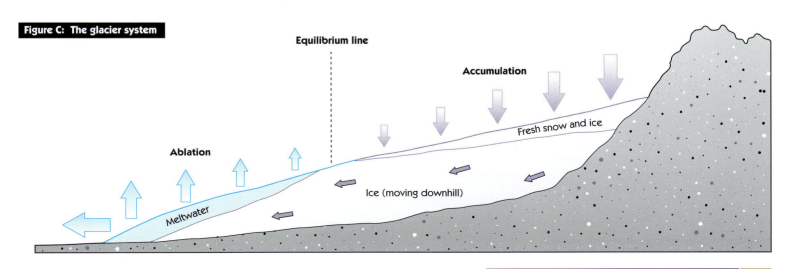

**Figure C: The glacier system**

**Figure A: The Rhône Valley, Switzerland**

## The glacier system

Figure C on page 79 shows the glacier system. The inputs to the glacier are snow and ice which build up in the zone of **accumulation**. The output of the glacier is water (melted snow and ice). The melting of a glacier is called **ablation**. Ablation is greatest in the summer and accumulation in the winter. If there is less ablation than accumulation the glacier's snout will advance and vice versa.

Figure A on this page shows the Rhône Glacier in Switzerland. The glacier is over 1 km wide and its length from the source to the snout on the photograph is 8 km. The Rhône Glacier's source is a huge corrie on the slopes of Dammastock, a mountain whose peak rises to 3630 metres above sea level. The glacier flows over the corrie lip at a height of 2800 m and descends the ice fall to the valley below. It flows down the valley until it

### ▼ Questions

**1** a How does snow become ice?
  b What causes ice to slide downhill?
  c Where do most glaciers have their sources?
**2** What is a glacier surge and what may be its cause?
**3** a What is ablation?
  b When during the year will ablation be at its maximum?
**4** Study the photograph of the Rhône Glacier and name the features at A, B, C and D.

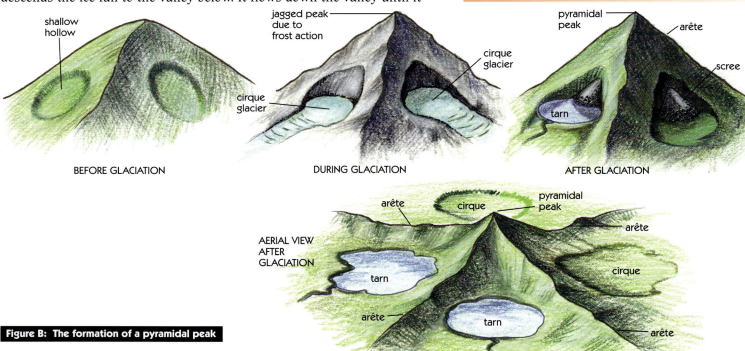

BEFORE GLACIATION — shallow hollow

DURING GLACIATION — jagged peak due to frost action, cirque glacier

AFTER GLACIATION — pyramidal peak, arête, scree, tarn

AERIAL VIEW AFTER GLACIATION — arête, cirque, pyramidal peak, arête, cirque, tarn, arête, tarn, arête

**Figure B: The formation of a pyramidal peak**

reaches 1800 m where the ice melts to form the River Rhône. The photograph was taken in 1925. Since then the glacier's snout has retreated and is found at a higher level, above 2100 m. This retreat is because of a slow increase in temperature since 1925.

## Glacial processes and landforms

### Erosion

Armed with fragments of rock called moraine, the glacial ice acts like a giant sheet of sandpaper, scratching and polishing the rock over which it passes. This process is called abrasion. Scratches called striations and smoothly polished surfaces called glacial pavements may be seen in many glaciated valleys.

### Arête

A corrie grows by erosion of the base and sides. Where two corries have formed beside each other, there may be only a very narrow ridge of land separating them. Such a ridge is called an arête.

### Pyramidal peak

Where several corries have been cut into a mountain from all sides they may form a pyramidal peak (Figure B). The Matterhorn in the Swiss Alps is a classic example.

### Glacial trough

A glacier can have great erosive power. The moving ice can carry rock fragments which scratch and wear away solid rock. The Rhône Glacier has deepened and widened the valley it flows along to create a glacial trough, or U-shaped valley (Figure C). This has steep sides and a broad valley floor. The interlocking spurs of the former river valley have been removed. The V-shaped cross-section of the river valley has been replaced by the U-shaped glacial trough.

Glacial troughs are not smooth. There are deeper rock basins, rock barriers and rock steps. After the ice has retreated, the rock basins may be occupied by long, narrow

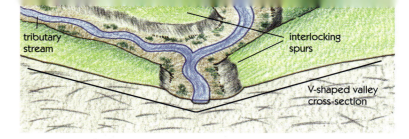

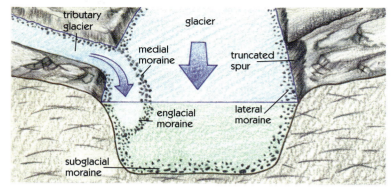

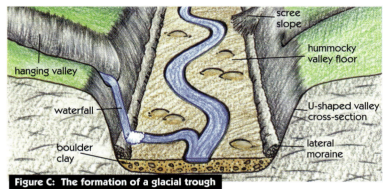

Figure C: The formation of a glacial trough

lakes called ribbon lakes. During the Ice Age the Rhône Glacier extended much further down the valley and Lake Geneva is an impressive example of a ribbon lake.

### Roche moutonnée

Where a more resistant outcrop of rock has been abraded by the ice a roche moutonnée is often formed (Figure D).

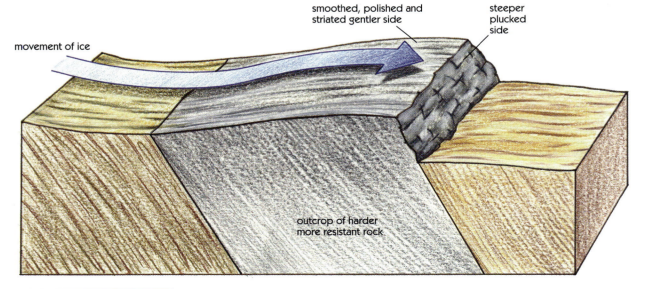

Figure D: The formation of a roche moutonnée

## Glacial transport

As a glacier moves across the landscape it collects rock fragments, or **moraine**. Some of these fragments fall on to the glacier from rocky slopes above them. Other fragments have been torn away from the valley floor and sides; such rock will have been weakened earlier by freeze-thaw action or pressure.

The glacier transports moraine in three ways (Figure E):
● lateral moraine is carried along the edges of the glacier (lateral moraine can be seen as a dark band along the edge of the Rhône Glacier in Figure A).
● englacial moraine is carried within the ice. It may have fallen down crevasses, been washed down by streams flowing across the ice, or have melted its way down into the glacier.
● ground moraine is moved along the bottom of the glacier.

Medial moraine is formed where two glaciers join and their lateral moraines combine.

## Glacial deposition

The load of a glacier will be dropped where the ice melts. This occurs mainly at the snout, but it can also occur underneath or at the sides of the glacier. Glacial drift is the general term given to these glacial deposits.

Glacial drift is of two types, depending on whether it has also been deposited by ice or meltwater. Till (also known as ground moraine or boulder clay) is the most common type of drift. It is deposited by ice. It is an unsorted jumble of jagged and angular rock fragments. These are very different from meltwater deposits. It is usually easy to tell whether particles have been deposited by ice or water because river deposits tend to be of roughly the same size at any one place: they are sorted. Also, the rocks deposited by rivers are rounded and smoothed by the water. River deposits are stratified (i.e. they are deposited in layers), each layer marking a particular period of deposition.

A number of distinctive landforms result from glacial deposition (Figure F).

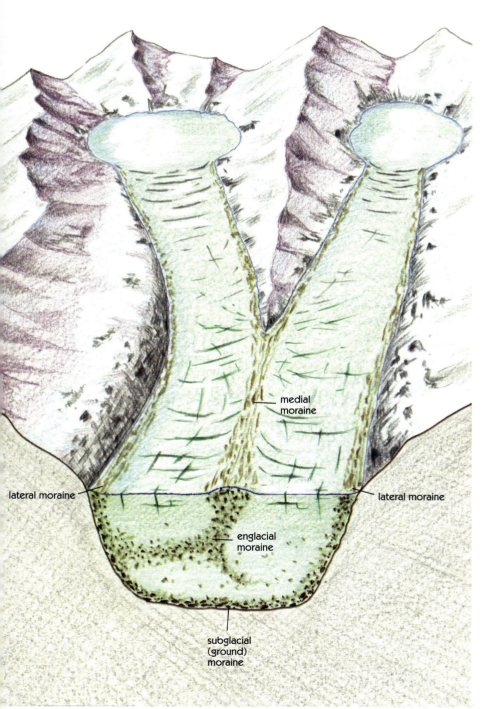

Figure E: How a glacier transports moraine

*Labels on figure:* medial moraine; lateral moraine; lateral moraine; englacial moraine; subglacial (ground) moraine

### ▼ Questions

**1** If you were walking along a valley in a glaciated highland region, what evidence would you look for to show that the valley had been formed by moving ice?

**2** Draw your own diagrams to explain the formation of
  a   corries
  b   glacial troughs.

**3** Study Figure G.
  a   Identify landforms A to E.
  b   Explain the difference between a terminal moraine and a recessional moraine.

**4** a   Explain what a drumlin is in your own words.
  b   How does a drumlin differ from a crag and tail? Illustrate your answer with two labelled diagrams.

## Terminal moraine

At the snout of the glacier (or ice sheet), moraine falls from the melting ice to form a ridge – called a terminal or end moraine. Many glaciated valleys have terminal moraines stretching across them, often now reduced to a line of hummocks dissected by streams. The terminal moraines of ice sheets are much larger landforms. The Cromer Ridge in Norfolk marks the limit of an ice sheet moving southwards across the North Sea. In places over 100m high, the Cromer Ridge includes boulders and stones from the nearby chalklands, as might be expected. More surprising are the rocks from the north of England, Scotland and even Norway which can be found within the Cromer Ridge. Such rocks have been carried by ice to an area far from their point of origin and are called erratics.

## Recessional moraine

The retreat of a glacier or ice sheet was rarely a smooth or rapid event. Usually the ice retreated in stages, sometimes with small advances as well. If the ice halted for long enough during its retreat, another terminal moraine might be formed. Such a moraine is called a recessional moraine; there may be several marked points along its line of retreat where the ice halted.

## Lateral moraine

In some glaciated valleys a terrace or moraine can be seen running along the valley side. This is the lateral moraine, formed at the sides of the glacier by rocks falling on to the ice from the slopes above. As the ice melted, the moraine subsided to its present position.

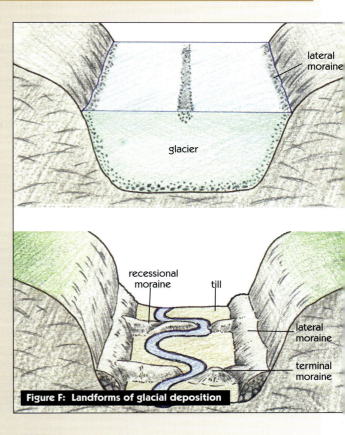

Figure F: Landforms of glacial deposition

## Drumlin

In the Eden Valley, east of the Lake District, are many low, rounded hillocks called drumlins. A drumlin lies parallel to the direction of ice flow with its steeper slope facing up-glacier. The drumlins in the Eden Valley are 10–50 m in height and 50–500 m in length. A group of drumlins is called a swarm and the resulting hummocky landscape is sometimes called 'basket of eggs' topography. Most drumlins consist only of boulder clay. Nobody is sure how drumlins are formed. It is thought that ground moraine has been shaped into drumlins by ice flowing over it. The streamlined shape of the drumlins reflects the wave-like movement of the ice.

## Crag and tail

Some drumlins have cores of solid rock. Their formation may be similar to the crag and tail found in Edinburgh. Castle Rock is a crag of very resistant basalt rock which withstood the passage of a glacier. In the lee of the crag a tail of weaker sedimentary rock has survived, protected from erosion by the ice. The Royal Mile runs along the gently sloping tail, leading up to the crag on which is perched Edinburgh Castle.

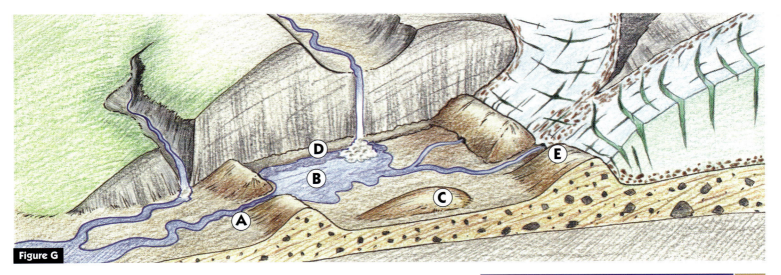

Figure G

The Lake District in north-west England was declared the country's first National Park in 1949. The National Park was established to preserve the natural beauty of this glaciated highland region. There is now no permanent ice in the Lake District, but during the Ice Age the region nourished its own ice cap. From the ice cap, glaciers flowed downwards and outwards, following the lines of least resistance – the existing pre-glacial river valleys which, in the Lake District, had formed a clear radial drainage pattern.

The effects of glaciation on the Lake District are revealed dramatically even today, over 10,000 years since the last glaciers vanished. The valley glaciers created impressive glacial toughs (Figure H shows an example in Hayeswater Gill). Following the disappearance of the glaciers, rock basins in the glacial troughs filled with water to form the many ribbon lakes which give the region its name.

The photographs and their captions (Figures A–H) show that there are fine examples to be found in the Lake District of most of the landforms of glacial erosion and deposition.

**Figure A:** Rigginsdale is an impressive glacial trough with a characteristic open U-shape cross-section. The peak to the right of the centre is Kidsty Pike, rising high over the small corrie, Sale Pot, at the head of Rigginsdale

**Figure B:** Red Tarn in winter. The snow emphasises the outline of this excellent example of a tarn

**Figure C:** The spectacular arête of Striding Edge. Winter sunshine picks out the knife-edge ridge, in places less than 50 cm wide with steep slopes falling over 300 metres on each side. The path along the arête descends from the summit of Helvellyn, from which the photograph was taken. It then rises to the peak of High Spying How shown in the centre of the photograph and then descends via Low Spying How and Bleaberry Crag. In the background the snow makes clear the impressive series of glaciated valleys, Grisedale, Deepdale and Patterdale. To the right rise the cliff-like slopes of Eagle Crag, rising above the neigbouring corrie, Nethermost Cove

**Figure D:** The Bowder Stone, an erratic rock which has been left precariously perched, is over 2000 tonnes in weight, showing that glacier ice can transport boulders of considerable size

**Figure E:** A roche moutonnée in Ennerdale. This mass of rock projects above the general level of the valley floor. The upstream side is clearly gently sloping while the downstream side is steeper, rougher and more jagged in appearance

**Figure F:** Dramatic screes of Wastwater, descending at an angle of over 35° for some 500 metres from the broken crags of Ilgill Head to the south-eastern shores of the lake

**Figure G:** Glacial till, or boulder clay, deposited as 'hummocky moraine' near the foot of Rossett Gill. Many of the glacial valleys are covered by such depositional features

**Figure H:** Thornthwaite Crag rising above Hayeswater Gill. The western side of the trough (on the right) includes an excellent example of a lateral moraine which the sunlight clearly illuminates

Our roads are choked with tourists' cars and coaches all summer long. I can't get to my fields, and I waste time when I try to drive into Keswick for supplies.

The mountains in winter are great for cross-country skiing, but some winters there isn't enough snow for downhill skiing.

I love living in the Lake District. The scenery is superb, the towns and villages are small and attractive. The quality of life here is so much higher than in the big cities.

We have to abide by strict planning regulations when we are constructing new roads and buildings. It's a bit of a hassle, to be honest.

There are climbs of every level of difficulty in the Lake District. It's a paradise for climbers.

There won't be any jobs for these two here. Hill farming's finished: the soil's poor and the slopes are too steep. Industry needs flat land for its buildings and room for expansion, and easy and quick communications. They won't find any of that here.

Although it is a beautiful place to live, and I suppose I'm lucky to have been born here, there really isn't a lot for young people to do. It's difficult to get around: all the roads have to follow the valleys and it can take ages to get anywhere! Sometimes I wish I could leap on one of the tourist coaches and go to live in Manchester.

**Figure I: Some of the advantages and disadvantages of living in an upland glaciated region such as the Lake District**

## Living in the Lake District

The Lake District is not only inhabited by tourists. Over 50,000 people live and work there. The economy of the region is highly dependent upon tourism, but farming, quarrying for slate and granite, forestry, water provision and some manufacturing industry all feature.

There are clear benefits and disadvantages in living in a glaciated highland region (Figure I). Tourists are attracted in great numbers: over a million visitors each year. This large number of visitors brings problems of congestion and pollution. Some of the more popular sites become 'honey pots', attracting a large share of the total visitors.

Some wealthier tourists purchase second homes in the Lake District to visit at weekends and holidays, and which for the rest of the year remain empty. Some of the more accessible pretty villages have so many second homes that village community life is badly affected. The demand for second homes pushes up house prices and makes it difficult for local people to afford houses in their own villages.

### ▼ Questions

1 Complete the table below.

| Glaciated Landform | Example(s) found in the Lake District |
| --- | --- |
| Corrie | Red Tarn |
| Arête | |
| Glacial trough | |
| Lateral moraine | |
| Erratic rock | |

2 List the advantages and disadvantages of living in a glaciated highland region shown in Figure I. Try to add as many others as you can.

### Review

Glaciers and ice sheets are major agents of erosion and deposition. Highland landscapes affected by glaciation have many characteristic landforms. Although the last glacial ice left the Lake District ten thousand years ago, the landforms of glacial erosion and deposition remain easily identifiable.

# Slopes

### Studying your local slopes

Are you aware of the slopes in your local area? If you live in a hilly district you certainly will be (Figure A). If you live in a lowland area you may not be so aware of the slopes. However, it is likely that the land is not completely flat. Even a very gentle slope can have an important effect upon the landscape especially in terms of water movement. Most landforms are made up of

slopes including larger features such as volcanoes, cliffs, escarpments and valleys, and smaller features such as river banks, drumlins and roche moutonnées. Some slopes near you may not be natural, for example railway cuttings, road embankments and waste tips.

To help in the description of slopes, names have been given to their separate parts, or elements (Figure B).

Figure A: Hilly landscape

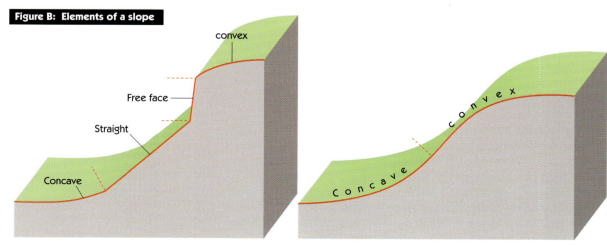

Figure B: Elements of a slope

convex

Free face

Straight

Concave

Concave    convex    convex

# What determines the shape of a slope?

- Rock type, for example strong resistant rocks such as granite or carboniferous limestone (Figure C) will have very different slopes from those developed in soft clay (Figure D). Complex slopes develop where more than one rock type is involved.
- Geological factors such as faulting and folding will affect the shape.

Past climates and processes such as glaciation and *periglaciation*.

- Present climate and processes.
- Vegetation.
- People's activities.

## How do slopes develop?

How will the slope in Figure C develop over time? In advanced level Geography there are several theories of slope development involving the decline, retreat or replacement of slopes. How the slope develops depends upon the processes operating on it, which may include:

- creep
- slide/fall
- flow.

Figure C: A carboniferous limestone slope

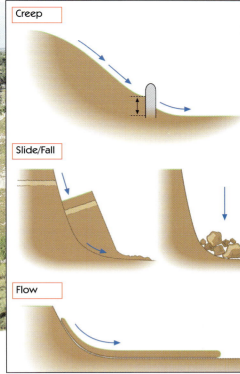

Creep

Slide/Fall

Flow

## ▼ Question

1 Which process or processes do you think will be operating on the slopes shown in Figure E?

Figure D: Clay slopes

Mesa

Figure E: Mesa, Landslide and Terracettes on the side of a chalk dry valley

Landslide

Terracettes

## Main activities

Drawing an annotated sketch and writing a guide book article.

## Key idea and question

● Landscapes in desert areas such as western USA are often spectacular and unique.
● What gives rise to such landscapes?

Monument Valley is located to the north-east of the Grand Canyon on the borders of the states of Arizona and Colorado in western USA (see Figure A). The *Lonely Planet* Guide gives the following information.

South West of Mexican Hat, Hwy 163 enters the Navajo Indian Reservation and Monument Valley. This is one of the most scenic drives in the Southwest, with sand dunes rolling up to a clear blue sky punctuated by sheer red buttes and colossal mesas – you've probably seen the landscape in a TV commercial or a Hollywood Western.

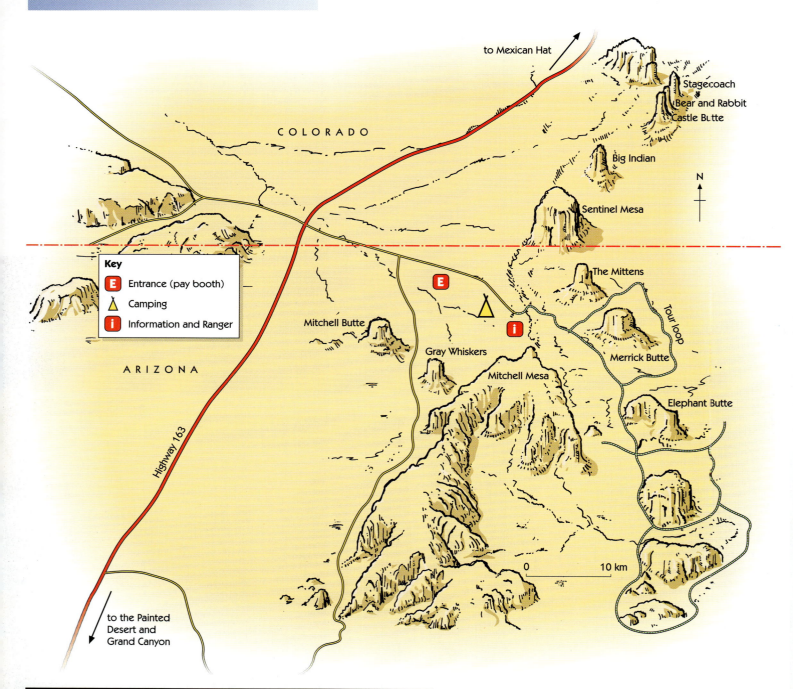

**Figure A: Monument Valley Navajo Tribal Path: main facilities and scenic attractions**

Figure B is a photograph of The Mittens which are two of the **buttes** in this desert landscape. These are flat-topped, steep-sided hills standing isolated on a flat plain. They are remnants of the much larger **mesas** which are steep-sided plateaux of rock. Both features are characterised by horizontally bedded rocks.

## How has this landscape been formed?

Although there is little water in these deserts now, many of the landforms have been formed over thousands of years by the action of water. When it does rain there is rapid run-off because of the lack of vegetation and deep soils. In winter temperatures are frequently below 0°C and **freeze-thaw** weathering beaks down the rocks. In summer the hot sun in the day can help to break down the rock surfaces (**exfoliation** weathering).

**Wind action** has also been involved, blowing away loose material to lower the surface. The wind also picks up sand and it wears away rocks; this 'sand-blasting' is called **abrasion**. The whole landscape in the Monument Valley has been uplifted over time and this has caused cracking and the weakening of the rocks. The rocks are 160 million years old and are sandstones which have different hardnesses. The harder rock on top of the buttes is like a hard cap which resists erosion. The softer lower sandstone, the De Chelly ('de-shay') sandstone erodes at different rates and helps to produce the spectacular scenery.

There are few landscapes of wind deposition in western USA but in the Sahara Desert sand dunes and other sand deposits make up 21% of the total landscape.

### Do you know?

**?** The three main types of weathering: Physical, chemical and biotic.
**?** It is the physical types of weathering that are important in Monument Valley

### ▼ Questions

**1** Draw an annotated sketch of the landscape shown in Figure B. Give your sketch a full title with a sub-title stating where the landscape is.
Annotate the following using four different colours:
  a  features of the rock e.g. different sandstones
  b  evidence of physical weathering
  c  the effects of water erosion
  d  the effects of wind erosion.
**2** Write a short article for a local guide book for the general public. Include a description of the landscape and explain how the landscape has been formed. ➡

### Review

Desert scenery is a result of local geology, physical weathering, and water and wind erosion.

**Figure B:** The Mittens in Monument Valley

Linking a map and photograph and drawing a fieldsketch. Writing a description and explanation.

## Key questions

● What are the distinctive landforms and slope developments in the Mendips?
● How and why have these features developed?

The Mendip Hills are in Somerset in south-west England. They rise to over 300 metres above sea level and appear dramatic as the surrounding land is an old sea bed which is very low lying and flat. To understand the slopes and landscapes of the Mendips, we need to know about the local geology.

During the Carboniferous Period about 300 million years ago, plants and sea creatures lived in this area in a warm sea. Their calcified remains sank to the bottom of the sea forming a thick layer of Carboniferous Limestone (Calcium carbonate). This layer covered the older Old Red Sandstone rock. The rocks were then folded into **anticlines** and **synclines** (referred to as rock structure). The limestone was gradually worn away from the anticline as shown in Figure B.

About 200 million years ago the sea covered the land again and new younger rocks were laid down only to be eroded away again. The Mendips developed as a flat topped plateau. During the last million years (the Pleistocene Period) ice reached as far south as the Mendips and left its influence on the landscape.

After each Ice Age there were warmer periods called the **interglacials**. It was in these times that meltwater began to cut deep river valleys, one of which was Cheddar Gorge (Figure E). As the water in the limestone was frozen the rock was no longer permeable: water could no longer pass through the joints and bedding planes. Figure C shows how the gorge became deeper with a new gorge being cut into the original gorge several times. One reason for the ledges has been the changing sea levels over time.

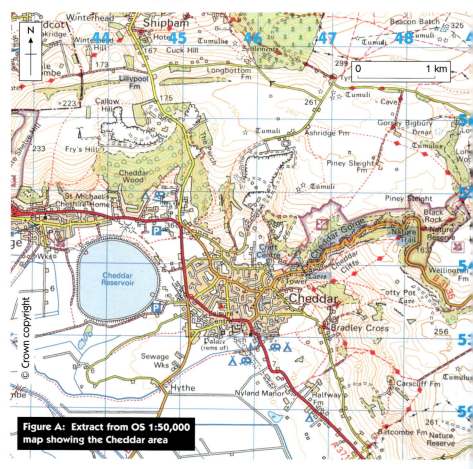

© Crown copyright

**Figure A: Extract from OS 1:50,000 map showing the Cheddar area**

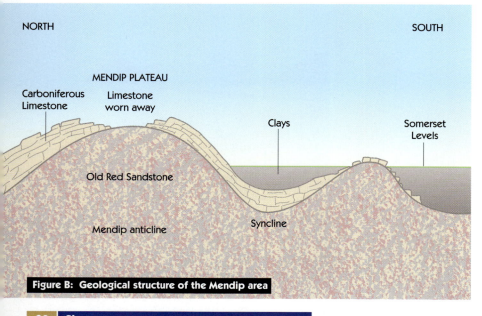

NORTH  SOUTH

MENDIP PLATEAU

Carboniferous Limestone
Limestone worn away
Clays
Somerset Levels

Old Red Sandstone

Mendip anticline
Syncline

**Figure B: Geological structure of the Mendip area**

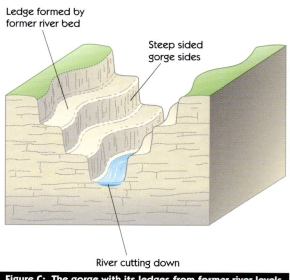

Ledge formed by former river bed

Steep sided gorge sides

River cutting down

**Figure C: The gorge with its ledges from former river levels**

When the ground was not frozen, floodwater was able to fill up the surface cracks of the limestone plateau. The acidic carbonic acid ($H_2CO_3$) produced from water and carbon dioxide then dissolved the Carboniferous Limestone ($CaCO_3$) carrying it along in the water as calcium bicarbonate ($Ca(HCO_3)_2$). This acidic water eroded the joints chemically, and the bedding planes made them larger and made some of them into cave systems.

Carboniferous Limestone is a very hard rock which is very permeable (but it is not porous). It has not smoothed over like the chalklands of England but has remained with steep slopes. Figure D compares the Carboniferous Limestone slopes with chalk slopes.

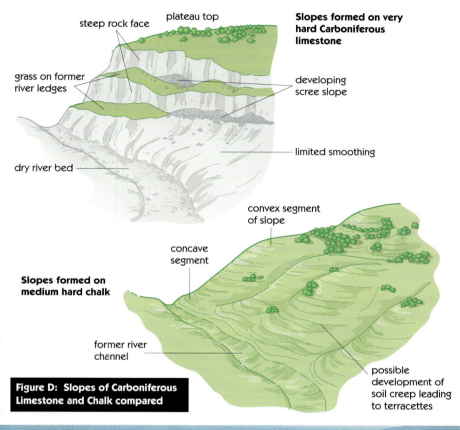

**Slopes formed on very hard Carboniferous limestone**

- plateau top
- steep rock face
- grass on former river ledges
- developing scree slope
- limited smoothing
- dry river bed

**Slopes formed on medium hard chalk**

- convex segment of slope
- concave segment
- former river channel
- possible development of soil creep leading to terracettes

**Figure D: Slopes of Carboniferous Limestone and Chalk compared**

**Figure E: Cheddar Gorge**

## ▼ Questions

1 Study Figures A and E.
  a Which way was the camera pointing in the photograph?
  b Name the reservoir in the distance.
  c What is the grid reference of the small lake shown in the photograph?
  d Of which river does this lake form a part?
  e Why is there no evidence of the river in the gorge itself?
  f Using only Figure A, note the evidence that the area is made of a hard rock such as limestone.

2 Draw a simple fieldsketch from Figure E. Label it to show the features mentioned in Question 1, including the features of the gorge itself.

3 Describe the landforms and slopes you can see in Figures A and E.

4 How were the landforms and slopes in the Mendip area affected by the following?
  a rock type
  b rock structures
  c the Ice Age and interglacials
  d erosion processes
  d present day weathering processes.

5 How do the slopes near Cheddar compare with those on the chalklands of England?

6 How do the slopes compare with another rock type you have studied?

# Summary questions

On this page are two photographs of UK landscapes showing distinctive landforms. Study the photographs and answer the questions below.

Figure A:  Coastline in Cornwall, showing Bedruthan Steps

Figure B:  Nant Ffrancon Valley, Snowdonia

## ▼ Questions

1 For each photograph:
   a  describe the landforms
   b  explain the processes that have created the landforms
   c  suggest how the landforms might change in the future.

2 For each photograph suggest how people may have used the areas shown and how such land use might have changed or interfered with the landforms.

| | | |
|---|---|---|
| **A** | Anemometer | Instrument used to measure wind speed. |
| | Anticline | An upward fold in rock strata. |
| **B** | Bankfull | Describes a river which is flowing in its channel up to the top of its banks. More discharge would lead to the river spilling over onto its *floodplain*. |
| | Barometer | Instrument used to measure atmospheric pressure. |
| | Biome | The mixed community of plants and animals (biotic community) occupying a major geographical region, e.g. tropical rainforest and tropical grassland and savanna. |
| | Botanically dry days | Completely dry days. Some parts of north-east Brazil experience 200–300 such days and only support drought-resistant savanna vegetation. |
| | Braided | Describes a river channel which is split into separate streams leaving islands. |
| **C** | Crust | The thin, solid, outermost layer of the Earth. It exists in two forms: the continental crust and the thinner, but denser, oceanic crust which is the floor of the ocean basins. |
| **D** | Deforestation | The clearance of forests. |
| | Desertification | The turning of land into desert, a process of land degradation in semi-arid lands where rainfall is unreliable. |
| | Discharge | The volume of water passing a point, obtained by multiplying the velocity of the river by the cross-sectional area. Discharge is measured in *cumecs*: cubic metres per second. |
| | Drainage basin | The total area drained by a river and all its tributaries. |
| **E** | Ecosystem | A system that shows the relationship between a community of living things (plants and animals) and their non-living environments. |
| | Exfoliation | The weathering action of intense daytime heat followed by cold night temperatures. The outer layers of rock can flake off, known also as 'onion skin' weathering. |
| **F** | Fault | A crack or fracture in a mass of rock along which there has been a movement of rock on one side relative to the other. |
| | Flood plain | The flat land forming the floor of a river valley. The area is regularly flooded by the river and is usually composed of sediment deposited by the river. |
| | Freeze-thaw | The weathering action of frost. As water freezes in rock cracks it expands by 10% creating weaknesses. Eventually particles break away from the main body of the rock. |
| **G** | Groyne | A barrier built on a beach to prevent the removal of the beach materials by longshore drift. It extends seawards, usually perpendicular to the coastline. |

| **H** | Harmattan | A hot dry wind blowing out from the Sahara affecting West Africa: a local name for the north-east trade winds. It affects more of the region in December–January but in June–July its influence is less as the region is under the influence of the low pressure. |
| | Horst | An uplifted block of rock between faults. It is also called a block mountain. |
| | Hydrograph | The discharge of a river plotted against time. Time can be a month, a year or a single storm; hence a *storm hydrograph*. |
| | Hygrometer | An instrument used to measure humidity. |
| **I** | Island arc | A chain of islands formed at a subduction zone as a result of the volcanism induced by the subducting oceanic plate. |
| | Isobar | A line drawn on a weather map through places which have equal atmospheric pressure. |
| **J** | Joint | A crack or fracture in a mass of rock along which there has been no movement of rock. |
| **M** | Magma | Molten rock lying under great pressure below the Earth's solid crust. It originates in the mantle. Magma is called lava when it erupts through a volcanic fissure. |
| | Mantle | The part of the Earth between the solid crust and the core. It is about 2900 km thick and is rich in silica. |
| **O** | Occluded front | Where the cold front catches up with the warm front, undercuts it and the warm sector of the depression is lifted above the Earth's surface. |
| **P** | Periglaciation | The processes associated with areas near to (peripheral to) ice sheets. Periglacial areas are areas of intense frost action and of *permafrost*. |
| | Permafrost | Permanently frozen subsoil. |
| **R** | Ribbon Lake | A long, deep, narrow lake in a glacially eroded valley e.g. Lake Maggiore and Lake Como in northern Italy, Lake Windermere and Ullswater in the English Lake District. |
| **S** | Sand dunes | An accumulation of sand deposited by the wind, as at the back of a beach or in desert areas. In desert areas such as the Sahara there can be many different shapes e.g. barchan dunes, linear dunes. |
| | Storm surge | The build-up of water in narrow sea areas by storm force winds. A storm surge may cause serious coastal flooding. |
| | Sustainability | The conservation of resources and environmental quality over time. |
| | Sustainable development | A form of use where resources are conserved over time. |
| | Syncline | Downward folded rock strata forming a trough. |
| | System | Simply a set of interrelated parts. Some systems are *closed* and others are *open*. All systems can be seen as having inputs, processes and outputs. |
| **W** | Watershed | The divide between two drainage basins; a ridge or an area of higher ground. |
| | Water table | The level within rock below which all the pores in the ground are saturated with water. Where the water table reaches the surface, areas of standing water exist. |